INTERNATIONAL UNION OF CRYSTALLOGRAPHY
CRYSTALLOGRAPHIC SYMPOSIA

INTERNATIONAL UNION OF CRYSTALLOGRAPHY
BOOK SERIES

This volume forms part of a series of books sponsored by the International Union of Crystallography (IUCr) and published by Oxford University Press. There are three IUCr series: IUCr Monographs on Crystallography, which are in-depth expositions of specialized topics in crystallography; IUCr Texts on Crystallography, which are more general works intended to make crystallographic insights available to a wider audience than the community of crystallographers themselves; and IUCr Crystallographic Symposia, which are essentially the edited proceedings of workshops or similar meetings supported by the IUCr.

IUCr Monographs on Crystallography

1 *Accurate molecular structures: Their determination and importance*
   A. Domenicano and I. Hargittai, *editors*

IUCr Texts on Crystallography

1 *The solid state: From superconductors to superalloys*
   A. Guinier and R. Jullien, *translated by* W. J. Duffin
2 *Fundamentals of crystallography*
   C. Giacovazzo, *editor*

IUCr Crystallographic Symposia

1 *Patterson and Pattersons: Fifty years of the Patterson function*
   J. P. Glusker, B. K. Patterson, and M. Rossi, *editors*
2 *Molecular structure: Chemical reactivity and biological activity*
   J. J. Stezowski, J. Huang, and M. Shao, *editors*
3 *Crystallographic computing 4: Techniques and new technologies*
   N. W. Isaacs and M. R. Taylor, *editors*
4 *Organic crystal chemistry*
   J. Garbarczyk and D. W. Jones, *editors*
5 *Crystallographic computing 5: From chemistry to biology*
   D. Moras, A. D. Podjarny, and J. C. Thierry, *editors*

Organic Crystal Chemistry

Papers presented at the
Seventh Symposium on Organic Crystal Chemistry
held at
Poznań-Rydzyna, Poland
14–17 August 1989

Edited by

J. B. Garbarczyk

Faculty of Chemical Technology
Poznań Technical University

and

Derry W. Jones

Chemistry and Chemical Technology
University of Bradford

INTERNATIONAL UNION OF CRYSTALLOGRAPHY
OXFORD UNIVERSITY PRESS
1991

CHEMISTRY

4657354

Oxford University Press, Walton Street, Oxford OX2 6DP
Oxford New York Toronto
Delhi Bombay Calcutta Madras Karachi
Petaling Jaya Singapore Hong Kong Tokyo
Nairobi Dar es Salaam Cape Town
Melbourne Auckland
and associated companies in
Berlin Ibadan

Oxford is a trade mark of Oxford University Press

Published in the United States
by Oxford University Press, New York

© *Contributors listed on pp. ix–xi, 1991*

A catalogue record for this book is available from the British Library

Library of Congress Cataloging in Publication Data
Symposium on Organic Crystal Chemistry (7th : 1989 : Poznań, Poland)
Organic crystal chemistry: papers presented at the seventh
Symposium on Organic Crystal Chemistry, held at Poznań-Rydzyna,
Poland, 14–17 August 1989 / edited by J. Garbarczyk and D. W. Jones.
p. cm. — (International Union of Crystallography
crystallographic symposia ; 4)
1. Crystallography—Congresses. 2. Chemistry, Organic—
Congresses. I. Garbarczyk, J. II. Jones, D. W. III. Title.
IV. Series.
QD901.S94 1989 548—dc20 91-16138

ISBN 0-19-855383-8

Printed in Great Britain by Bookcraft Ltd.,
Midsomer Norton, Avon

Preface

Structural crystallographic studies can determine not only the full stereochemistry of chemical species but also details of their arrangement in the crystal. Such geometrical data provide an essential basis for the interpretation of chemical, physical, and biological properties of chemical species.

Since 1977, the Symposia on Organic Crystal Chemistry, organized by Professor Z. Katuski from the Faculty of Chemistry, Adam Mickiewicz University at roughly two-yearly intervals, have provided a forum for substantial international participation within a meeting of predominantly Polish chemical crystallographers. The overall aim of these meetings, and of this volume, is not only to illustrate recent developments and results in molecular structure determination but also to discuss the relevance of this information to the chemistry and biology of the compounds studied.

At the Seventh Symposium, held in the congenial surroundings of Rydzyna chateau, near Poznań, in August 1989, eighty lectures, talks and posters were presented to nearly two hundred participants. Among the themes discussed were factors influencing molecular conformation and polymorphism, chemical and biological activity, intermolecular interactions, crystal chemistry of polymers and molecular modelling.

The formal versions of the contributions included in this volume comprise a cross-section of contemporary organic crystal chemistry, as summarised in the Introduction. A short final chapter draws on a Round Table Discussion at the close of the Symposium and attempts to look forward to organic crystal chemistry in the 1990's.

Poznań   J.B.G.
Bradford   D.W.J.
July 1991

Contents

Contents

Contributors

L. Addadi
Department of Structural Chemistry
Weizmann Institute of Science
76100 Rehovot
Israel

Andrew R. Barron
Department of Chemistry
Harvard University
Cambridge
MA 02138
USA

Joel Bernstein
Department of Chemistry
Faculty of Natural Sciences
P.O.B. 653
84105 Beer Sheva
Israel

R. Boese
Institut für Anorganische Chemie
   der Universität-GH Essen
Universitätstr. 5–7
D-4300 Essen 1
Germany

D. Bläser
Institut für Anorganische Chemie
   der Universität-GH Essen
Universitätstr. 5–7
D-4300 Essen 1
Germany

C. E. Briant
Chemistry and Chemical Technology
University of Bradford
West Yorkshire BD7 1DP
England

W. L. Duax
Medical Foundation of Buffalo Inc.
   Research Institute
73, High Street
Buffalo, NY 14203-1196
USA

J. L. Finney
Rutherford Appleton Laboratory
Chilton, Didcot
Oxon OX11 0QX
England

J. Garbarczyk
Faculty of Chemical Technology
Technical University of Poznań
Pl-60-965 Poznań
Poland

J. F. Griffin
Medical Foundation of Buffalo Inc.
   Research Institute
73, High Street
Buffalo, NY 14203-1196
USA

O. Johnson
Chemistry and Chemical
   Technology
University of Bradford
West Yorkshire BD7 1DP
England

D. W. Jones
Chemistry and Chemical
   Technology
University of Bradford
West Yorkshire BD7 1DP
England

A. Kálmán
Central Research Institute for Chemistry
Hungarian Academy of Sciences
Budapest
P.O.B. 17, H-1525
Hungary

Andrzej Katrusiak
Crystal Chemistry Laboratory
Faculty of Chemistry
Adam Mickiewicz University
Grunwaldzka 6
60-780 Poznań
Poland

J. Kowalik
Faculty of Chemistry
Jagiellonian University
Karasia 3
30-060 Kraków
Poland

Tadeusz Marek Krygowski
Department of Chemistry
University of Warsaw
ul. Pasteura 1
02 093 Warsaw
Poland

Leslie Leiserowitz
Department of Structural Chemistry
Weizmann Institute of Science
76100 Rehovot
Israel

M. Lahav
Department of Structural Chemistry
Weizmann Institute of Science
76100 Rehovot
Israel

Janusz Lipkowski
Institute of Physical Chemistry of
    the Polish Academy of Sciences
ul. Kasparzaka 44/52
01-224 Warszawa
Poland

T. Miebach
Institut für Anorganische Chemie
    der Universität-GH Essen
Universitätstr. 5–7
D-4300 Essen 1
Germany

N. Niederprum
Institut für Anorganische Chemie
    der Universität-GH Essen
Universitätstr. 5–7
D-4300 Essen 1
Germany

B. J. Oleksyn
Faculty of Chemistry
Jagiellonian University
Karasia 3
30-060 Kraków
Poland

D. Paukszta
Faculty of Chemical Technology
Technical University of Poznań
Pl-60-965 Poznań
Poland

P. Serda
Regional Laboratory of Physico-
    chemical Analysis and Structural
    Research
Jagiellonian University
Karasia 3
30-060 Kraków
Poland

J. D. Shaw
Chemistry and Chemical
  Technology
University of Bradford
West Yorkshire BD7 1DP
England

L. J. W. Shimon
Department of Structural Chemistry
Weizmann Institute of Science
76100 Rehovot
Israel

J. Śliwiński
Faculty of Chemistry
Jagiellonian University
Karasia 3
30-060 Kraków
Poland

T. Sterzynski
Faculty of Chemical Technology
Technical University of Poznań
Pl-60-965 Poznań
Poland

M. Vaida
Department of Structural Chemistry
Weizmann Institute of Science
76100 Rehovot
Israel

C. C. Wilson
Rutherford Appleton Laboratory
Chilton, Didcot
Oxon OX11 0QX
England

1

Introduction: current topics in organic crystal chemistry

D. W. Jones

Unlike many physical methods, for which early applications to organic chemistry have been semi-empirical, X-ray crystallography has long been a major source of specific information about the geometry of the whole molecule inferred or determined by diffraction from the electron density pattern in crystals. Later developments in collection and treatment of X-ray data for structure solution and refinement enabled this approach to be extended to finer details of the electron distribution and of thermal motion of atoms and groups and to be applicable rapidly to larger organic structures. Also, the distinctive contribution of neutron measurements, derived by diffraction by a different atomic scattering mechanism, has been demonstrated more widely, for example in precise location of hydrogen atoms and for combination with X-ray data. Concurrently there have been advances in the complementary approach to calculation by molecular mechanics and other methods of the expected stereochemistry and dimensions of free molecules.

From the 1930's onwards, experimental determinations of molecular shapes and dimensions in organic crystal chemistry have formed a major application of X-ray crystallography. Examples range from the single-crystal determination of a particular three-dimensional molecular structure in an organic compound to much more general considerations of polymorphism, intermolecular bonding, strain, lattice forces, etc., which, in some cases, involve a survey of the wealth of data now available from very many compounds. It is these wider organic solid state topics that largely feature in the present volume, as summarized below.

Polymorphism, whether arising merely from differences of molecular arrangement or also involving differences of molecular conformations, is widespread among organic solids. Since polymorphic organic systems can involve one molecular species crystallizing with different geometrical arrangements of both the molecules themselves and also their mode of packing, they provide a rewarding means of studying structure-property relations and the influence of crystal forces on

2

D. W. Jones

molecular conformation. In his survey of polymorphism, Bernstein illustrates the relevance of conformational polymorphism (when intramolecular interactions are small) to molecular properties through a combination of X-ray crystallographic and ultraviolet-visible spectroscopic studies of the benzylideneanilines. When intramolecular interactions are not negligible, polymorphism can aid the investigation of bulk properties, as with planar organic conductors or the problems of aggregation and J-bonding in organic dyes.

In a review of the clathrate-forming properties of co-ordination complexes $M^{II}X_2A_4$ (where, typically, M is a transition or other metal ion, X may be CN^-, Cl^-, NO_2^-, etc., and A is a pyridine base), Lipkowski focuses on guest-lattice interactions of Werner clathrates. Clathrate-forming ability is influenced by the molecular shape, packing, isomerism, and flexibility of the host complex. $Ni(NCS)_2(4\text{-methylpyridine})_4$ provides an example of a versatile and hospitable host whose structure adapts, when forming inclusion compounds, to the steric requirements of the guest. In general the Werner clathrates are interpreted as interstitial solid solutions of the guest in a metastable polymorph structure of the host.

Duax and Griffin tackle the physiologically important problem of the mutual selectivity between receptors and steroid hormones, anti-hormones, chemicals and drugs through the wealth of crystallographic data available on over 400 steroids. These help one to predict preferred conformations and the ability of steroid hormones to interact with a receptor. As a result of surveys of receptor binding and activities, Duax and Griffin have been able to model receptor interactions of steroid hormones and so deduce probable bindings. For estrogen and progestin receptor sites, an A-ring binding/D-ring active model is proposed, whereas for androgen receptors a D-ring binding/A-ring active model appears more likely.

Crystal structure determinations provide an expanding fund of data potentially available for the investigation of inter- and intramolecular interactions. Not only soft structural parameters, such as hydrogen bonds and interatomic separations in complexes, but also hard parameters - bond lengths and angles - may be affected slightly by packing forces in crystals. Krygowski has applied a simple force-field approach to estimate geometrical deformations brought about by intermolecular interactions in simple aromatic systems such as para-nitrobenzoate anions and (to illustrate appreciable intramolecular interactions) the 2,6- and 3,5-dimethyl derivatives of N,N-dimethyl-para-nitroaniline.

Solvent can exert a marked influence on the growth and dissolution and hence on the morphology, structure and symmetry of molecular crystals. Attempts to elucidate at the molecular level the role of solvent-surface interactions have been concerned either with reduction of interfacial tension on specific faces so that a roughened interface encourages growth or with a raised energy barrier to the removal of bound solvent molecules that have been preferentially adsorbed.

Leiserowitz and colleagues have made two stereochemical approaches to study the effect of solvent binding, and its analogy with additive binding, on crystal growth and dissolution.   In one, the growth of some crystalline hydrates in the presence of molecularly analogous (or "tailored") solvents showed growth inhibition by stereospecific solvent adsorption at a particular face.   In the other approach, relative rates of growth and dissolution of hemihedral faces at opposite ends (donor and acceptor) of polar axes were explained in terms of differential solvent binding or repulsion between two kinds of surface site.   It was concluded that desolvation of the tailored solvent from a crystal surface can be the rate-determining step for growth at the face. Differential binding of a normal solvent at a surface may encourage a repetitive sequence of solvent binding at some sites, solute adsorption at others, followed by solvent expulsion and fast growth.

In another study of intermolecular interactions in hydrogen-bonded crystals, Katrusiak describes structural transformations induced by high hydrostatic pressures on simple cyclic β-diketoalkanes.   Comparisons between the effects of compression on the 1,3-cyclopentane- and -hexanediones (which undergo phase transitions) and 2-methyl-1,3-cyclopentanedione (which has anti-parallel hydrogen-bonded molecular chains and is almost unaffected) point to the significance of electrostatic interactions affected by chain polarity.

Aromatic molecules into which strain is introduced by the annulation of a small alicyclic ring, as in cyclobutabenzene, are often thermally labile or are liquid at room temperature.   To overcome consequent problems of crystallization and single-crystal X-ray diffractometry, Boese describes an apparatus for low-temperature crystallization whereby seed formation is reduced but growth facilitated and the selected crystals transferred to the diffractometer.   Bond lengths and angles and difference electron-density distributions have been determined accurately at around 110 K for eight strained compounds from crystals prepared by the above techniques.   In an alternative approach to the Mills-Nixon effect, the concept of bent bonds appears to be at least qualitatively valuable for interpreting bond lengths unless highly electron-withdrawing groups are present.

Many polycyclic aromatic hydrocarbons (PAH) are not quite planar in their crystal structures;   substituents can increase the deviation from planarity, as, for example, with the isomeric monomethylbenz-[a]anthracenes.   Briant, Johnson, Jones and Shaw have examined the bond lengths and angles associated with $-CH_3$ and $-OCH_3$ substitution in phenanthrenes, benz[a]anthracenes, benzfluoranthenes and other PAH in the Cambridge Crystallographic Database.   They note that PAH methoxy groups commonly orientate almost perpendicular to the ring, and sometimes lie in the molecular plane, but rarely have intermediate torsion angles.

The contributions of p and d orbitals to sulphur-oxygen bonding at tetracoordinated sulphur have been the subject of extensive theoretical discussion and bond-length measurements in both organic and inorganic chemistry.   In a short article dedicated to the late J.I. Musher, who predicted the existence of diacycloxyspirosulphuranes, Kálmán discusses the continuity between long hypervalent S → O bonds and short S...O intramolecular contacts in cyclic sulphuranes and recently synthesised sulphonium salts.   Recent crystal structure analyses provide examples both of unsymmetrical three-centre-four-electron O-S$^+$...O= and N-S$^+$...O= systems (N-S$^+$, O-S$^+$ ∼ 1.7Å; S$^+$...O ∼ 2.3Å) and genuine (Musher-type) hypervalent O-S-O (and N-S-N) nearly linear bridges in symmetrical spirosulphuranes incorporating aromatic rings.

Most of the detailed information about molecular structures in solids has come from X-ray diffraction on single crystals.   Neutron diffraction, in which the scattering is by the nucleus (and hence is isotope-dependent) rather than by electrons, is particularly advantageous for hydrogen (or deuterium) atom location in organic crystals and when samples must be enclosed in environmental chambers and for dynamical studies.   However, in addition to high cost and limited availability, continuous neutron beams from nuclear reactor sources, even those of so-called "high flux", are of lower effective intensity than beams from X-ray tubes, partly because much of the beam is wasted in wavelength selection.   Generation of a pulsed neutron beam by spallation of pulses of accelerated protons from a U or Ta target, as at the ISIS source, RAL, England, enables much more efficient use to be made of the neutrons by time-of-flight techniques.   Finney & Wilson list these advantages and draw particular attention to the extensive and well resolved data that can be derived at this most powerful source from powder samples of, for example, squaric acid ($C_4H_2O_4$) and benzene, where powder profile refinement can yield structural parameters of virtually single-crystal quality.   Single-crystal diffraction and spectroscopic applications of the spallation pulsed neutron source are also outlined.

There follow two examples of the utilization of X-ray crystal structure determinations in the study of influences on the stereochemistry of reactions.   The bulky oxygenphilic reagent methylaluminium 2,6-di-tert-butyl-4-methylphenoxide (MAD) selectively activates organocarbonyl moieties and so provides a means of C-C bond formation.   In the light of structural characterization of carbonyl adducts of MAD, Barron concludes that the aryloxide oxygen-aluminium π-bonding influences the preferred orientation of the aryloxide ligands and so controls the stereoselectivity.

For quinine (Q) and analogous Cinchona alkaloids, either only the aliphatic tertiary-amine nitrogen N(1) is protonated, to form quininium cation QH$^+$, or both N(1) and the aromatic quinoline N(3) are protonated to form QH$_2^{2+}$.   The exact conformation of these QH$^+$ cations and their hydrogen bonding and other interactions with anions and water may be relevant to biological blocking reactions.   Accordingly, Oleksyn, Śliwiński, Kowalik and Serda present a stereochemical survey of cation-anion interactions, with special reference to torsion angles involving C(9), which links the two parts of the cation, and N(1).

The contribution by Paukszta, Garbarczyk and Sterzynski returns to the theme of polymorphism, here the influence of electric fields on the formation and transformations of modifications of synthetic polymers. It is found that a strong (~ 1 kV) electric field reduces the amount of hexagonal (β) isotactic polypropylene and, together with temperature, accelerates the transformation to the monoclinic (α) form.

2

Polymorphism and the investigation of structure–property relations in organic solids

Joel Bernstein

1. INTRODUCTION

It is axiomatic that structure plays an important role in determining the properties of materials. Understanding that role is crucial in our efforts to design materials with predetermined properties. In seeking to understand the relationship between structure and properties we attempt to design experiments which will allow us to eliminate as many variables as possible and to focus strictly on the structural variations and their manifestations in the variation of chemical and/or physical properties. The existence of polymorphic forms provides a unique opportunity for the investigation of structure-property relationships, since the only variable between two polymorphic crystal forms is that of structure. In such cases variations in properties must be due to differences in structure. and the analysis and interpretation of these experiments is considerably facilitated by the limited number of variables.

Polymorphism in organic solids, until recently considered a curiosity by many, has proven to be a fairly widespread phenomenon. Nearly 3000 entries in the January '89 release of the Cambridge Structural Database contain qualifying information which indicates the existence of polymorphic forms for the compound in question. Thus, there already exists in the literature a plethora of suitable subjects for the study of a wide variety of phenomena, and the number of investigations employing polymorphic systems has been growing steadily in recent years.

When considering polymorphic systems for the study of structure-property relations, organic crystals may be classified into two broad classes. In the first, the crystal is considered to be an 'oriented gas', and one may consider the crystal structure as simply a matrix which holds non-interacting or weakly interacting molecules in a particular spacial arrangement. Such cases provide opportunities for the investigation of molecular properties. When the different polymorphic structures also display different conformations of the same molecule (i.e *conformational polymorphism*), there is a unique opportunity to study, say, the changes in spectroscopic or electronic properties as a function of variation in molecular conformation and the influences of the crystalline environment on molecular conformation. In the second class of materials, the molecules are considered to be strongly interacting and suitable experiments then yield information on the relationship between structure and the properties of the bulk (e.g., conductivity, magnetic properties, etc.).

Over the past few years a variety of techniques, including crystallographic, computational and spectroscopic, have been employed to study polymorphic systems. In this summary we briefly survey some of the physical and chemical properties which have been studied on polymorphic systems.

Many of the examples of polymorphism and techniques for investigating and utilizing it come from those areas of chemical research where full characterization of a material is crucial in determining its ultimate use, for instance in the pharmaceutical (Haleblian and McCrone, 1969; Haleblian, 1975; Clements, 1976), dye (Walker *et al*,1972; Griffiths and Monahan, 1976; Etter *et al*, 1984; Tristani-Kendra *et al*, 1983; Morel *et al*, 1984), and explosives (Karpowicz *et al*, 1983) industries. Various aspects of the subject have been treated in books (Varna and Krishna, 1966; Byrn, 1983; Kuhnert-Brandstatter, 1971). The occurrence of the phenomenon of polymorphism is still not widely recognized although it has been the subject of a number of reviews (Haleblian and McCrone, 1969; Haleblian, 1975; McCrone, 1963).

2. INVESTIGATION OF MOLECULAR PROPERTIES

Let us examine first the utilization of polymorphism to study molecular properties. When a molecule with one or more conformational degrees of freedom crystallizes in polymorphic structures, there is a possibility for the occurrence of *conformational polymorphism*: the existence of different molecular conformations in the various crystal structures (Bernstein, 1987). Conformational polymorphs may be expected to exist when conformational energy minima differ by less than roughly 2 kcal-mol^{-1} and the energy barriers between those minima are sufficiently low to allow the simultaneous existence of a number of conformations in solution. Under such circumstances slightly different crystallization conditions such as temperature or rate of evaporation (or in some instances identical conditions) lead to different polymorphs with different conformations.

In a system exhibiting conformational polymorphism the molecular moiety, of course, is constant among the crystal structures. Thus there are two structural variables: the molecular conformation and the crystal packing. Understanding the relationship between these two structural variables can provide information on the relationship between crystal structure and molecular conformation, which in turn is related to the validity of using information on molecular conformation derived from solid state studies to interpret various phenomenon in other phases of matter. In a more general sense, the study of the phenomenon of conformational polymorphism also provides information on the environmental influences on molecular conformation. In addition, different molecular conformations are characterized by different electronic structures, and the latter may be manifested in the spectral properties, which may then be expected to vary among the polymorphs.

A nearly ideal system for investigating both of these aspects of conformational polymorphism has been that of the benzylideneanilines **I**. The molecule has only

I (X,Y = Cl, Br, CH$_3$)

two conformational degrees of freedom (denoted α and β). A study of the molecular energetics for X=Y=H shows that a minimum is obtained that for α=45°, with an energy *ca* 1 kcal-mol^{-1} below the planar conformation (α=0°) (Bernstein, Hagler and Engel, 1981). The potential energy function for β rises from a minimum at β=0°. Substitution at the 4 and 4' positions is not expected to significantly perturb these energy profiles. For the paradisusbstituted derivatives of **I** there are nine possible isomers containing various combinations of Cl, Br and CH$_3$ as substituents. All of these have been prepared and their crystal structures have been determined (Bernstein and Schmidt, 1972; Bernstein and Izak, 1975, 1976; Bernstein, Bar and Christensen, 1976; Bar and Bernstein, 1977, 1982b, 1983, 1987).

Polymorphs have been identified and studied in two of the homodisubstituted derivatives ($X=Y=CH_3$; $X=Y=Cl$). The former case is trimorphic while the latter is dimorphic; both exhibit conformational polymorphism. In such cases the relationship between the molecular conformation and the crystal structure may be studied by a variety of computational methods. The molecular energetics, relating the energies of the molecular conformations found in the various crystal structures, have been determined by *ab initio* methods. For larger systems the techniques of molecular mechanics are quite suitable as well. For cases such as those found here, in which the various conformers correspond to different energies, the environment of the molecule in the crystal (*i.e.* the crystal structure) must provide sufficient energy to stabilize the otherwise high energy conformation. The lattice energies may be determined by energy calculations based on the atom-atom potential method. (Kitaigorodsky and Mirskaya, 1972; Mirsky, 1976, 1980; Mirsky and Cohen, 1976; Williams, 1972, 1981; Williams and Starr, 1977; Ramdas and Thomas, 1978; Kitaigorodsky, 1978).

In both the dichloro and the dimethyl systems one of the polymorphs is a crystal structure containing an essentially planar molecule, which is more energetic than the most stable, non-planar conformation by approximately the same 1 kcal-mol^{-1} noted above. In the dimethyl system the molecule in one of the three polymorphs does exhibit the most stable conformation, while in the dichloro system the second polymorph has a molecule with an intermediate conformation (due to disorder) ($\alpha=25°$; $\beta=-25°$). In both of these systems the calculations indicate that the lattice energy for the structure of the more highly energetic molecule is sufficient to stabilize those otherwise unfavorable conformations (Bernstein and Hagler, 1978; Bar and Bernstein, 1982a).

Is it possible to attribute the stabilisation of one structure relative to another to any specific types of interactions or to interactions between particular pairs of atoms or regions of the molecule? This question can be answered by partitioning the total energy into the individual atomic contributions of which it is composed. Comparison of the atomic contributions for the various structures then reveals the differences and similarities among the structures. For these two cases at least, and we suspect that for virtually all crystal structures for which the intermolecular interactions are those of the van der Waals type, the differences in lattice energy between polymorphs are due to a cooperative effect. Comparison of the contribution of an atom in one structure to its counterpart in another structure consistently shows that essentially all of the atoms contribute by a small amount to the stabilization of the more stable form. Additional details of the computational treatment of conformational polymorphs can be found in a recent review (Bernstein, 1987).

Clearly in these examples there are significant differences in the molecular conformations among the various polymorphs, even though only two conformational parameters are involved. Because these conformational parameters involve linkages between π-electron systems, even relatively small values for α and β can lead to significant changes in the electron energy levels in the molecule. These changes should then be manifest in the UV-Vis spectra of the molecules in the various polymorphs.

Benzylideneaniline **I** is isoelectronic with stilbene **II** and azobenzene **III**.

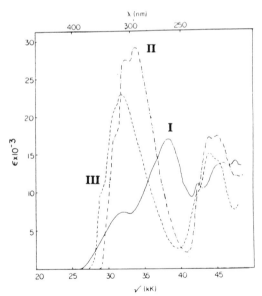

Therefore, it might be reasonable to expect that the solution absorption spectra of these three molecules (Figure 1) would be

Figure 1. Solution absorption spectra of benzylideneaniline **I**, stilbene **II**, and azobenzene **III**.

very similar to each other. In fact, it is seen that the spectrum of **I** differs considerably from that of **II** and **III**, which indeed are very similar to each other. This difference in spectral response aroused the interest of spectroscopists quite a while ago (Haselbach and Heilbronner, 1968, and references therein). It was attributed to a non-planar conformation for **I**, while **II** and **III** were believed to be planar, at least on the average, in solution. We have already seen that the planar conformation for **I** is not the most stable one, so this interpretation comes as no surprise. Perhaps it is well to note here at least the qualitative reasons for the lack of planarity of **I**.

All three compounds contain three π-electron systems, so that conjugation among these π systems is a driving force to planarity. For a planar conformation of **I** the distance between the hydrogen on the central bridge and the *ortho* hydrogen on the aniline ring is very short, being slightly less than 2.0Å. This creates considerable steric strain, which can be relieved by a rotation about the N-phenyl bond, thus breaking the conjugation over the whole molecule. The short intramolecular H...H distance is obviously absent in **III**, and in **II** the H...H distance is increased due to the longer C=C bond compared to the C=N bond in **I** (Burgi and Dunitz, 1971).

The existence of the two polymorphic derivatives of **I** (X=Y=Cl) allows us to measure directly the differences in spectral response which result from differences in molecular conformation. Since we are interested in comparing *molecular* properties we must assume that the 'oriented gas' model is valid here, i.e. there are no significant interactions between molecules, and that the various crystal structures merely serve to hold the molecules in a particular conformation.

Clearly, we must measure the ultraviolet and visible absorption spectral properties of the molecules in the crystal. Highly absorbing substances would normally demand the preparation of extremely thin samples of material. However, the use of polarized nearly-normal incidence reflection spectroscopy on single crystals not only overcomes this problem but provides some additional advantages as well (Anex, 1966; Pennelly and Eckhardt, 1976). First, the use of polarized light allows one to gain information on the directional properties of the electronic transitions as well information on the energies and intensities. Second, the fact that measurements are made on crystal surfaces rather than through the bulk allows one to choose a crystal face which provides a suitable projection of the molecules for study of a particular property. The availability of a number of faces for investigation due to the three dimensional nature of the crystal also provides for redundancy for confirmation of results.

We now describe the application of this technique to the polymorphic dichlorobenzylideneanilines (Eckhardt and Bernstein, 1972; Bernstein, Anderson and Eckhardt, 1979). For both structures the projections and the suitable faces chosen for study are given in Figure 2. Since the lowest energy transition in stilbene and azobenzene is long-axis polarized, we would like to choose crystal faces which give good projections of the long axis of the molecule. It is seen that this is the case for the (001) face chosen for the triclinic form and the (010) face chosen for the orthorhombic form. The principal directions are also given on this figure. The two spectra for each face were measured with the light polarized parallel to these two directions.*

The reflection spectra (Figure 3) clearly exhibit differences between the two polymorphs. The spectrum for the triclinic structure (Figure 3a) which is polarized very nearly along the long

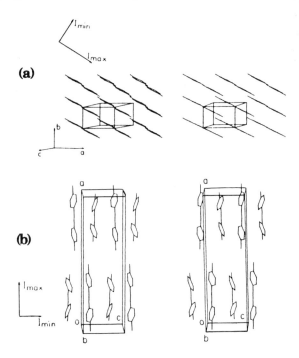

Figure 2. Projections of the crystal structures of **I** (X=Y=Cl) onto the faces studied spectroscopically. (a) The (001) face of the triclinic form. (b) (110) face of the orthorhombic structure. I_{min} and I_{max} are the extinction directions which were oriented parallel to the electric vector of the incident light for the measurement of the reflection spectra.

---

*In the orthorhombic structure the principal directions correspond to symmetry axes. Due to the absence of symmetry in the triclinic structrure they may exhibit dispersion with wavelength, but none was observed in this case.

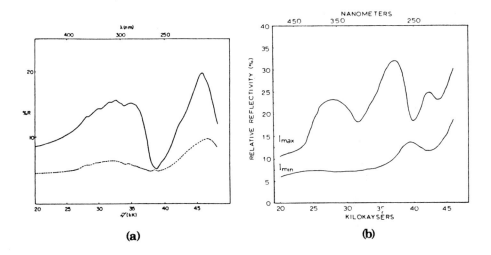

(a) **(b)**

Figure 3. Reflection spectra at 298 K for the two forms of **I** (X=Y=Cl). (a) The (001) face of the triclinic structure; (——) and (·····) are the spectra for light polarized parallel to the I_{max} and I_{min} directions respectively. (b) The (110) face of the orthorhombic form.

axis of the molecule exhibits the single band characteristic of stilbene and azobenzene, while the spectrum for orthorhombic form polarized along essentially along the same molecular axis (see Figure 3b) is clearly different. These reflection spectra may be converted to the corresponding absorption spectra via the Kramers-Kronig transform (Anex, 1966; Anex and Fratini, 1964)*. The results, shown in Figure 4, may be considered as the absorption spectra which would be obtained if it were possible to measure them through the particular face specified.

As expected, the absorption spectra exhibit the same differences as the reflection spectra. The spectrum of the planar molecule in the triclinic structure not only exhibits a single band, but this band is punctuated by reproducible vibronic structure due to the superposition of the C=N stretching mode (ca 1350 cm^{-1}) on the long axis electronic transition. Similar structure (due to the N=N and C=C stretching modes, respectively) can be seen in the azobenzene and stilbene solution spectra in Figure 1 (Dyck and McClure, 1962). In the non-planar molecular conformation found

_____

*The relationship between the reflection and absorption spectra is the same as that between optical rotatory dispersion and circular dichroism.

in the orthorhombic structure, the conjugation of the π system is broken, so the long axis transition no longer exists, and consequently no vibronic structure is expected or observed.

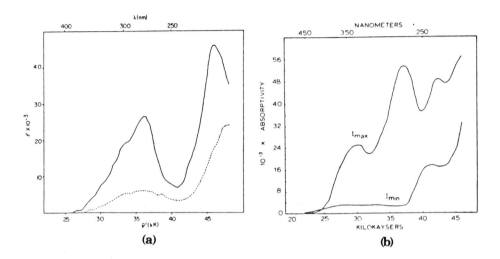

Figure 4. Kramers-Kronig transformed spectra of **I**, obtained from the reflection spectra in Figure 3. The captions for (a) and (b) are as in Figure 3.

It is important to reiterate here the nature of this experiment. A single molecular species is forced to adopt different molecular conformations as a result of its crystallizing in two different polymorphic crystal structures, The resulting crystal structures then serve as matrices to hold these molecules in the different conformations to permit the investigation of the manifestations of different molecular structure on the electronic energy. The energies of the various bands observed spectroscopically in the planar and non-planar conformations compare very favorably with the values determined from photoelectron spectroscopy (Bally *et al*, 1976). This fact, coupled with the observation that the spectrum for the orthorhombic material very closely resembles the solution spectrum, vindicates the initial assumption that these crystals behave as oriented gases. This experiment clearly demonstrates the utility of utilizing conformational polymorphs to study the relationship between molecular conformation and electronic structure.

3. INVESTIGATION OF BULK PROPERTIES

3. 1. Electrical Conductivity

We turn now to a discussion of the importance of polymorphism for the study of properties of the bulk, as opposed to the molecular properties just described. Bulk properties of organic materials depend on the interactions between molecules as well as the intrinsic nature of the molecules. One area of a great deal of activity in recent years has been in the field of electrically conducting organic materials (Wudl, 1984; Williams *et al*, 1985). In order to design and prepare substances which will conduct electricity it is first necessary to understand the relationship between structure and conductivity for these materials (Shaik, 1982). A series of studies involving a polymorphic system provided confirmation of one of the important structural criteria for conductivity.

Virtually all of the organic conductors known are based on planar or nearly planar π donors and/or acceptors. The first known organic conductor was the complex between the π donor tetrathiafulvalene **IV** and the π acceptor tetracyanoquinodimethan **V** (Ferraris *et al*, 1973; Coleman *et al*, 1973). A distinguishing

IV   **V**

structural feature of the structure of this complex was the presence of segregated stacks of donors and acceptors rather than mixed stacks of donors and acceptors which is characteristic of the vast majority of π electron donor acceptor (EDA) complexes (see Figure 5).

Since that first discovery, much of the progress in this field has involved primarily variations on the theme of the donor **IV**, or the acceptor **V**. One of the variations on the donor is tetramethyl-tetraselenofulvalene **VI**, first prepared by Bechgaard *et al* (1980,

VI

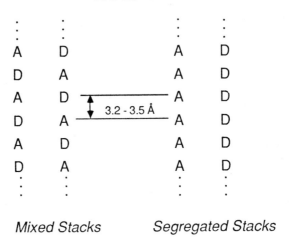

Mixed Stacks   Segregated Stacks

Figure 5. Schematic representation of the stacking modes of donors and acceptors in charge transfer complexes.

1981). **VI** forms a pair of polymorphic 1:1 complexes with **V**. The structures (Bechgaard *et al*, 1977; Kistenmacher *et al*, 1982) (Figure 6) show very clearly the presence of mixed stacks in one form and segregated stacks in the other. The mixed stack structure is a semiconductor while the segregated stack structure is a conductor. While the presence of segregated stacks was considered important for electrical conductivity in these materials, the study of this polymorphic system demonstrates that it is indeed a necessary condition.

Obviously, then, to prepare conducting materials of this type, it is necessary to have control over the mode of crystallization - to be able to prepare the segregated stack structures. This polymorphic system also provides some very useful information in that regard. First, the preponderance of mixed stack structures of π EDA complexes suggests that the mixed stack motif is thermodynamically preferred over the segregated stack motif (Shaik, 1982). If such is the case, how is it possible to obtain the energetically less preferred but structurally desirable segregated stack structure? The answer lies in the methods used to crystallize the two forms. Both forms of the complex of **VI** and **V** crystallize from acetonitrile, but the crystallization technique is different (Bechgaard, 1977). For the mixed stack complex contained in transparent, red crystals, crystallization is carried out at nearly equilibrium conditions, as might be expected for the thermodynamically more stable form: solutions of donor and acceptor are mixed and allowed to evaporate slowly. The opaque

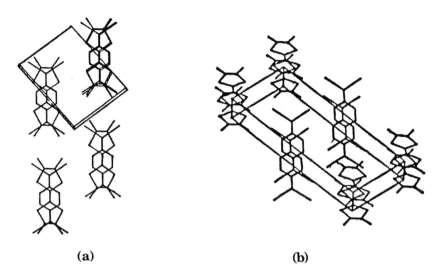

(a) **(b)**

Figure 6. Views of the two polymorphic structures of TMTSF:TCNQ (**VI:V**). In both cases the view is on the plane of the TCNQ molecule.(a) The mixed stack complex. (b) Segregated stack structure.

shiny black crystals of the second polymorph, typical of conductors and composed of segregated stacks, are obtained by what might be called kinetic, rather than equilibrium, conditions: hot saturated solutions of donor and acceptor are mixed and rapidly cooled, resulting in the formation of microcrystals. These are removed and used as seeds for the growth of larger crystals of this form, this time under equilibrium conditions. The importance of these observations on crystallization is twofold. First, it provides rather compelling experimental evidence about the energetic relationships between these two structural classes of EDA complexes. Second, knowing that, it suggests a recipe for obtaining the higher energy, but in many cases more desirable polymorphic form.

3.2. Aggregation of Dyes

Some of the riddles of the bulk spectroscopic properties of organic crystals, as opposed to those of individual molecules described earlier, have also been clarified as the result of studies on polymorphic systems. Over fifty years ago Jelley (1936) discovered a phenomenon which is now recognized as rather widespread among organic dyes. He found that, when the solution concentration of many dyes is increased, the intensity of the characteristic molecular absorption band decreases at the expense of the growth of an intense, narrow absorption band which is red-shifted with respect to the molecular absorption band (see Figure

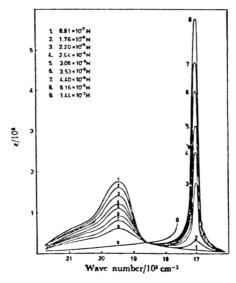

Figure 7. Solution spectra of a dye exhibiting J-banding.

7). The new absorption, called a J-band after Jelley, is commonly attributed to the formation of aggregates of dye molecules (Smith, 1974; Herz, 1974)). While the structure and properties of these aggregates is still the subject of investigation, they have proved extremely useful in the photographic sciences. The aggregates adsorb onto silver halide crystals and photographic scientists have taken advantage of the phenomenon for many years to tune the spectral response of photographic film (Smith, 1974).

An understanding of the phenomenon of J-banding requires knowledge of the nature of the aggregates, in order to relate spectral properties to structure. Direct methods for studying the structure are still not available, but the crystal structure of a dye can certainly be used as an approximation of the structure of the aggregates. Unfortunately, many of the dyes that would be suitable candidates for such studies exhibited a considerable reluctance to form crystals which would be suitable for crystallographic studies. Recently, however, the crystal structures of an increasing number of dyes have been published, so that much more structural data are available for relating spectral properties to structure (Marchetti, Salzberg and Walker, 1976; Tanaka, Tanaka and Hayakawa, 1980 Dulmage, *et al*, 1978).

Again assuming that the crystal structure provides a reasonable model for the structure of a dye aggregate, a polymorphic system would provide a particularly suitable system for studying the relationship between structure and spectral properties of dyes. One polymorphic system is the squarylium dye **VII**, which crystallizes in a triclinic form with violet colored crystals and a monoclinic form with green colored crystals (Bernstein and Chosen, 1988). This particular

VII

compound has been touted as potential photovoltaic material (Morel,1979; Forster and Hester, 1982; Morel *et al*, 1978), because of its photovoltaic conversion efficiency approaching 25%.

The crystal structures may be readily compared in Figure 8. The triclinic structure requires that all molecules be parallel as can be readily seen in the Figure. The monoclinic structure has an additional element of symmetry in the typical herring-bone pattern so there are two sets of parallel molecules. This is essentially the simplest perturbation that can be made to the triclinic structure, and one would expect a difference in the spectral properties corresponding to this structural perturbation, if they depend on the geometrical relationship between neighboring molecules. The polarized normal incidence reflection spectra are given in Figure 9, and it can be readily seen that the spectra are significantly different between the two forms for light polarized nearly along the long axis of the molecule. This must be a consequence of the difference in the interaction of a single molecule with its surroundings. The reflection spectrum of the band in the monoclinic form has been interpreted as being comprised of two oscillators, while that of the triclinic form contains three, or possibly four, oscillators (Tristani-Kendra and Eckhardt, 1984; Tristani-Kendra, Delaney and Eckhardt, 1985).

The difference in properties between forms might make one more desirable in terms of its use, which again raises the question as to how the two forms are obtained. In fact in this case crystals of both forms appear simultaneously in the same crystallizing vessel. This suggests that they are energetically very similar, in spite of the apparent differences exhibited in Figure 8. A more detailed look at the crystal structures in a different manner is very

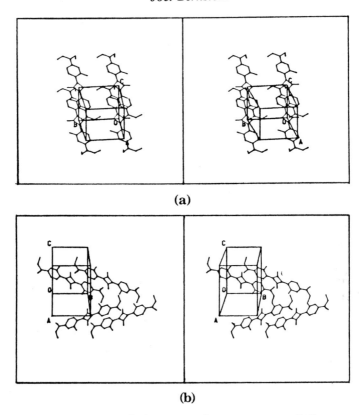

(a)

(b)

Figure 8. Stereoviews of the crystal structures of the squarylium dye **VII**. (a)   Triclinic structure, (100) face. (b) Monoclinic structure, (100) face.

revealing. Instead of projecting the structure on the crystal face as in Figure 8, we project one molecule onto the plane of the closest neighbor. The views thus obtained for the two forms are compared in Figure 10. It is now seen that there is a striking similarity between the two in terms of this relationship of neighboring molecules. This suggests that such a stacking interaction is a dominant one for this type of molecule. In order for both of the crystals to form the stacking interaction must exist in solution. Its presence in both of the structures also suggests that this is a good model for at least one type of aggregate of the material. Then the similarities and differences between the two structures may be looked at in a slightly different, but perhaps more understandable way.   Both structures contain these plane-to-plane stacks. In the triclinic form they are simply lined up parallel to one another; in the monoclinic form this parallel alignment of stacks also occurs, but it is interspersed with columns which are related to each other by a screw axis. The fact that the two crystals appear simultaneously is simply evidence for that fact that the screw

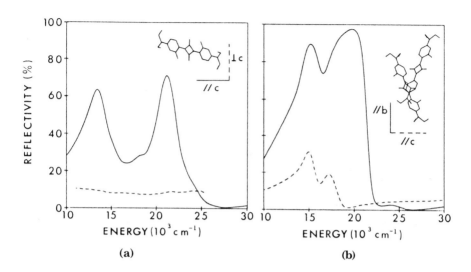

Figure 9. Normal incidence polarized reflection spectra of the two forms of **VII**. In each case there are two spectra measured with the light polarized along each of the two directions, as indicated in the upper right hand corner, which also shows the projection of the molecule(s) onto the crystal face studied. (a) Triclinic polymorph, (100) face. (b) Monoclinic polymorph, (100) face.

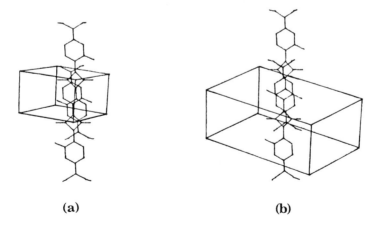

Figure 10. Molecular overlap in each of the two structures of **VII**. (a) Triclinic structure. (b) Monoclinic structure.

relationship is a very weak one energetically; it must be prevented to obtain a dominance of the triclinic structure and it must be sustained to obtain a dominance of the monoclinic structure.

4. CONCLUDING REMARKS

The distinguishing feature of polymorphic systems of organic compounds is the constancy of the molecular species while geometric features vary. Thus differences in properties between polymorphs can be attributed exclusively to geometric factors. The relationship between structure and properties is thus directly determined through the study of polymorphic systems. We have presented here a number of examples of studies which take advantage of the existence of polymorphism to study the physical and chemical properties in organic crystals. The existence of conformational polymorphism allows the study of the influence of crystal forces on molecular conformation. Properties of molecules which depend strongly on conformation may also be investigated directly utilizing the conformational polymorphs. Bulk properties of crystals, which depend on the spatial relationship between neighboring molecules, are particularly amenable to such investigations. Such investigations are not limited to the known polymorphic systems. Many polymorphs have been discovered serendipitously; in fact there are those that claim that virtually every material will prove to be polymorphic, if sufficient effort is expended on it (McCrone, 1963). Chemists are in the business of making materials which do things better or in a different way. Many of the structure-property relations on which those strategies are based are still not understood. Considerable understanding can come from the utilization of polymorphic systems using methods similar to those described above.

REFERENCES

ANEX, B.G. (1966). Molec. Cryst. 1, 1

ANEX, B.G. and FRATINI, A.V. (1964). J. Mol Spectry 14, 1

BALLY, T. HASELBACH, E., LANYIOVA, S., MARDCHER, F. and ROSSI, M. (1976). Helv. Chim. Acta 59, 486

BAR, I. and BERNSTEIN, J. (1976). Acta Crystallogr. B32, 1609

BAR, I. and BERNSTEIN, J. (1977). Acta Crystallogr. B33, 1738

BAR, I. and BERNSTEIN, J. (1982a). J. Phys. Chem. 86, 3223

BAR, I. and BERNSTEIN, J. (1982b). Acta Crystallogr. B38, 121

BAR, I. and BERNSTEIN, J. (1983). Acta Crystallogr. B39, 266

BAR, I. and BERNSTEIN, J. (1987) Tetrahedron 43, 1299

BECHGAARD, K., CARNEIRO, K., RASMUSSEN, F.B., OLSEN, RINDORF, M.G. JACOBSEN, C.S., PEDERSEN, H.J. and SCOTT, J.C. (1981). J. Am. Chem. Soc. 103, 2440

BECHGAARD, K., JACOBSEN, C.S., MORTENSEN, K., PEDERSEN, H.J. and TORUP, N. (1980). Solid State Commun. 33, 1119.

BECHGAARD, K., KISTENMACHER, T.J., BLOCH, A.N. AND COWAN, D.O. (1977). Acta Crystallogr. B33, 417

BERNSTEIN, J. (1987) in G. DESIRAJU ed., Organic Solid State Chemistry, Elsevier, Amsterdam, pp. 471-518

BERNSTEIN, J., ANDERSON, T.E. and ECKHARDT, C.J. (1979). J. Am. Chem. Soc. 101, 541

BERNSTEIN, J. and CHOSEN, E. (11988). Mol. Cryst. Liq. Cryst. 164, 213

BERNSTEIN, J. and HAGLER, A.T. (1978). J. Am. Chem. Soc. 100, 673

BERNSTEIN, J. HAGLER, A.T., and ENGEL, M. (1981). J. Chem. Phys. 75, 2346

BERNSTEIN, J. and IZAK, I. (1975). J.Cryst. Mol. Struct. 5, 257

BERNSTEIN, J. and IZAK, I. (1976). J. Chem. Soc. Perkin II, 1976

BERNSTEIN, J. and SCHMIDT, G.M.J. (1972). J. Chem. Soc. Perkin II, 951

BEVERIDGE, D.L., and JAFFE, H.H. (1966). J. Am. Chem. Soc. 88, 1948

BURGI, H.-B. and DUNITZ, J.D. (1971). Helv. Chim. Acta 54, 1255

BYRN S.R. (1983). Solid State Chemistry of Drugs, Academic Press, New York

CLEMENTS, J.A. (1976). Proc. Analyt. Div. Chem. Soc., 21

COLEMAN, L.B., COHEN, M.J., SANDMAN, D.J., YAMAGISHI, F.G., GARITO, A.F. and HEEGER, A.J. (1973). Solid State Commun. 12, 1125.

DULMAGE, W.J., LIGHT, W.A., MARINO, S.J., SALZBERG, C.D., SMITH, D.L. and STAUDENMEYR, W.J. (1978). J. Appl. Phys. 49, 5543

DYCK, R.H. and McCLURE, D.S. (1962). J. Chem. Phys. 36, 2326

ECKHARDT, C. J. and BERNSTEIN, J. (1972). J. Am. Chem. Soc. 94, 3247

ETTER, M.C., KRESS, R.B., BERNSTEIN, J. and CASH, D.C. (1984).
J. Am. Chem. Soc., 106 6921

FERRARIS, J., COWAN, D.O., WALATKA, V. and PERLSTEIN, J.H. (1973). J. Am. Chem. Soc. 95, 948

FORSTER, M. and HESTER, E. (1982). J. Chem. Soc. Faraday. Trans. I, 78, 1847

GRIFFITHS, C. H. and MONAHAN, A.R. (1976). Molec. Cryst. Liq. Cryst., 33, 175

HALEBLIAN, J.K.and McCRONE, W.C. (1969). J. Pharm. Sci., 58, 411

HALEBLIAN, J.K. (1975). J. Pharm. Sci., 64, 1269

HASELBACH, E. and HEILBRONNER, E. (1968). Helv. Chim. Acta 51, 16

HERZ, A. (1974). Photgr. Sci. Eng. 18, 441

JELLEY, E.E. (1936). Nature 138, 1009

KARPOWICZ, R J., SERGIO, S.T. and BRILL, T.B. (1983). I&EC Prod. Res. & Dev. 22, 363

KISTENMACHER, T.J., EMGE, T.J., BLOCH, A.N. and COWAN, D.O. (1982), Acta Crystallogr. B38, 1193

KITAIGORODSKY, A.I. (1978). Chem. Soc. Revs. 7, 133

KITAIGORODSKY, A.I. and MIRSKAYA, K. (1972). Mat. Res. Bull. 7, 1271

KUHNERT-BRANDSTATTER, M. (1971). Thermomicroscopy in the Analysis of Pharmaceuticals , PergamonPress, New York

MARCHETTI, A.P., SALZBERG, C.D. and WALKER, E.I.P. (1976). J. Chem. Phys. 64, 4693

McCRONE, W.C. (1963) in FOX, D., LABES, M.M. and WEISSBERGER, A. (Eds.), Physics and Chemistry of the Organic Solid State, Vol. I, Interscience, New York, p. 725

MIRSKY, K. (1976). Acta Crystallogr. A32, 199

MIRSKY, K. (1980). Chem. Phys. 46, 445

MIRSKY, K. and COHEN, M.D. (1976). J. Chem. Soc. Faraday II 72, 2155

MOREL, D.L. (1979). Molec. Cryst. Liq. Cryst. 50, 127

MOREL, D.L., GHOSH, A.K., FENG, T., STOGRYN, E.L., SHAW, P.E. and FISHMAN, C. (1978). Appl. Phys. Lett. 32, 495

MOREL, D.L., STOGRYN, E.L., GHOSH, A.K., FENG, T., PURWIN, P.E., SHAW. R.F., FUSHMAN, C., BIRD, G.R. and PIECHOWSKI, A.P. (1984). J. Phys. Chem., 88, 923

PENNELLY, R.R. and ECKHARDT, C.J. (1976). Chem Phys 12, 89

RAMDAS, S. and THOMAS, J.M. (1978). Chem. Phys. Solids and Surfaces 7, 31

SHAIK, S.S. (1982). J. Am. Chem. Soc. 104, 5328

SMITH, D. (1974). Phot. Sci. Eng. 18, 309

TANAKA, J., TANAKA, M. and HAYAKAWA, M. (1980). Bull. Chem. Soc. Jpn 53, 3109.

TRISTANI-KENDRA, M., DELANEY, J. and ECKHARDT, C.J. (1985). J. de Physique 46, C7 425

TRISTANI-KENDRA, M. and ECKHARDT, C.J. (1984). J. Chem. Phys. 81, 1160

TRISTANI-KENDRA, M., ECKHARDT, C.J., GOLDSTEIN, E. and
BERNSTEIN, J. (1983). Chem. Phys. Lett., 98, 57

VARNA, A.R. and KRISHNA, P. (1966). Polymorphism and Polytypism inCrystals, J. Wiley and Sons, New York

WALKER, M.S. MILLER, R.L., GIFFITHS, C.H.and GOLDSTEIN, P. (1972). Molec. Cryst. Liq. Cryst.,16, 203

WILLIAMS, D.E. (1972). Acta Crystallogr. A28, 629

WILLIAMS, D.E. (1981). Topics in Appl. Phys., 26, 3

WILLIAMS, D.E. and STARR, T.L. (1977). Comp. Chem. 1, 173

WILLIAMS, J.M., BENO, M.A., WANG, H.H., LEUNG, P.C.W., ENGE, Y.J., GEISER, U. and CARLSON, K.D. (1985). Acc. Chem. Res. 18, 261

WUDL. F. (1984). Accts. Chem. Res. 17, 227

3

Werner clathrates: guest–lattice intermolecular interactions

Janusz Lipkowski

Introduction

The 'clathratogenic' ability of coordination complexes of the general formula $M(II)X_2A_4$ was first reported by Schaeffer and coworkers [W.D. Schaeffer et al., J. Am. Chem. Soc., 79(1957), 5870]. During 1950-1970 the main reason for studying the subject was the possibility of practical use of clathration selectivity in separation of mixtures which are otherwise separable with difficulty, i.e. mixtures of isomeric compounds. Thus, the early literature contains extensive information on selectivity of clathration of a variety of mixtures of organic compounds (guests) by different Werner complexes (host) {Table 1}.

Table 1

CLATHRATE FORMING WERNER COMPLEXES (HOST) $M^{II}X_2A_4$

W. S. Schaeffer et al.

M^{II} = $Fe^{2+}, Co^{2+}, Ni^{2+}, Zn^{2+}, Cd^{2+}, Mn^{2+}, Hg^{2+}, Cr^{2+}, ...$

X = $NCS^-, NCO^-, CN^-, NO_3^-, NO_2^-, Cl^-, Br^-, I^-, ...$

A = Pyridine bases (39 listed in the literature), isoquinoline

P. de Radzitzky, J. Hanotier et al.

M^{II} = Ni^{2+}

X = NCS^-

A = α-Alkylarylamines (47 listed in the literature)

More recently, the accumulation of basic structural and physicochemical knowledge on clathrate compounds [J. Lipkowski, in J.L. Atwood, J.E.D. Davies and D.D. MacNicol (Eds.), 'Inclusion Compounds', Acad. Press (1984), London, vol. 1, ch. 3, and J. Hanotier and P. de Radzitzky, idem, vol. 1, ch. 4] enabled their nature to be rationalized better. It is worth noting here that the original definition of the compounds as 'clathrates' could be based merely on the observation that there is no apparent reason for any chemical bonding between the host and guest components, and selectivity may be correlated with steric rather than chemical properties of the guests. The present review concentrates on the structure of Werner clathrates and on the guest-lattice intermolecular interactions. It is subdivided into four parts in which the versatility of Werner complexes in clathrate formation is presented at the level of:

(i)   the host-complex chemical composition,
(ii) the molecular structure of the host complex (i.e. its isomerism and possibility of conformational change),
(iii) the crystalline structure of the host (i.e. its polymorphism), and
(iv) the guest-lattice interactions within a given clathrate structure. In the final section an attempt is made to classify Werner clathrates as a particular class of molecular crystals - interstitial solid solutions of the guest in a metastable (or unstable) crystalline host.

I. Chemical composition of the host.

Although there is no quantitative measure of 'clathratogenic ability' it may be assumed that this property of Werner clathrates is correlated with molecular shape of the host complex which, in turn, is primarily defined by its chemical composition. MeX_2A_4 complexes having unsubstituted pyridine as the 'A' ligand do not, reportedly, form clathrate inclusion compounds. Introducing a methyl group in position 3- or, preferably, 4-, however, is enough to obtain highly versatile hosts in clathrate formation. It seems that molecular packing of the non-clathrated crystal structure of the host complex is of importance: complexes which form close packed structures in pure form show little, if any, tendency to co-crystallize with foreign (guest) molecules.

From the data in Table 1 it is clear that there is a large variety of complexes of different chemical composition. The list is by no means complete and, moreover, the possibility of 'mixed' complexes, i.e. containing more than one type of ligand 'A', must be considered, as shown by D.R. Bond et al. [S. Afr. J. Chem. 36(1982),19] (Fig. 1).

Fig. 1. Molecular structure of bis(isothiocyanato)-bis(4-methylpyridine)-bis (4-phenylpyridine) nickel II complex.

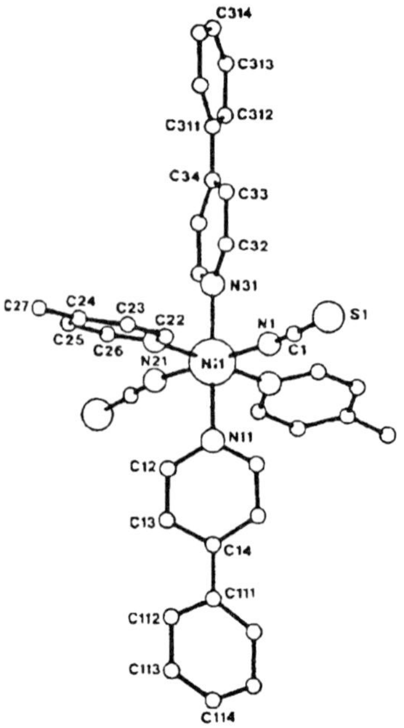

Fig. 2. The crystal structure of bis(isothio-cyanato)-tetrakis(alpha-phenylethylamine) nickel(II) complex (non-clathrated form) viewed along b axis.

Fig. 3. The crystal structure of bis(isothio-cyanato)-tetrakis(alpha-phenylethylamine) nickel(II) s-buthylbenzene clathrate viewed along a axis.

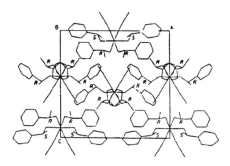

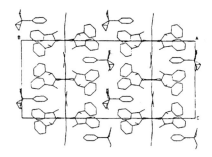

II. Molecular structure of the host complex

Since the clathrate-forming ability of a given host complex is related to its molecular shape, then clathration must be influenced by isomerism and conformational flexibility of the host molecule. The first has been clearly demonstrated by Nassimbeni et al. [J. Chem. Soc., Chem. Commun., 1985, 1788] in the bis(isothiocyanato)-tetrakis(alpha-phenylethylamine) nic-kel(II) complex which, in pure form, crystallizes as cis-coordinated (with respect to isothio-cyanates) (Fig. 2) while the same host material co-crystallized with s-buthylbenzene (guest) contains trans-coordinated complex molecules (Fig. 3)

The conformational flexibility of a host complex depends on its chemical structure. In the simplest case of 4-methylpyridine complexes (Fig.4), stable conformations are limited to two different types in which pyridine moieties form a four-blade 'propeller' (Fig. 4) having either all four pyridine rings twisted in the same direction with respect to the equatorial plane of

Fig. 4. An ORTEP projection of the bis(isothiocyanato)-tetrakis(4-methyl pyridine) nickel II complex molecule.

Fig. 5. A projection of the bis(isothiocyanato)-te-trakis(4-methylpyridine) copper II complex molecule.

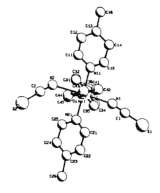

the complex (the '++++' conformational type) or, less energetically favourable [J. Lipkowski, J. Mol. Struc. 75 (1981), 13] conformation of the '++--' type. The anionic ligands may also contribute to the conformational flexibility of the host. In Fig. 5, which shows a projection of the $Cu(NCS)_2$(4-Methylpyridine)$_4$ complex molecule (as found in a clathrate structure of the complex [J. Lipkowski, A. Kislykh and Yu.A. Dyadin, J. Incl. Phenom., submitted]), the bond angle and conformation of the anionic group are evidently significant

Fig. 6. Structure of the bis(isothiocyanato)-tetrakis(4-ethylpyridine) nickel II complex molecule.

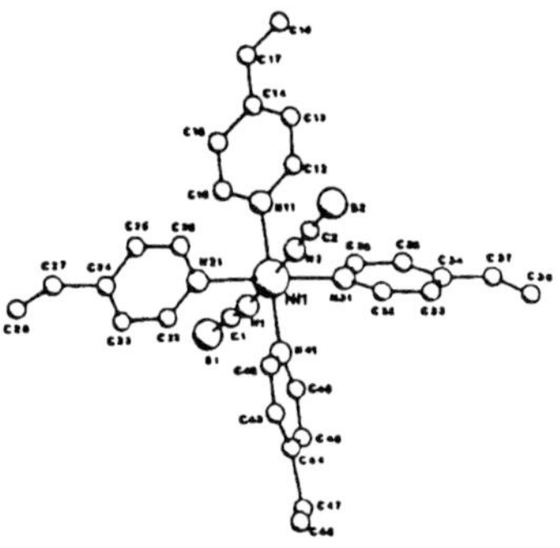

Fig. 7. Structure of the bis(isothiocyanato)-tetrakis(4-phenylpyridine) nickel II complex molecule.

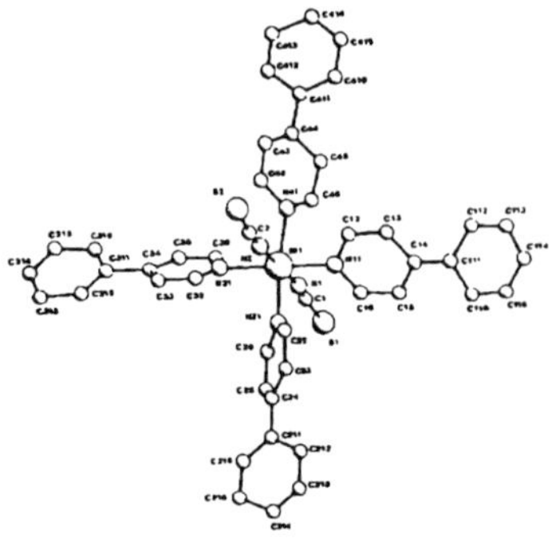

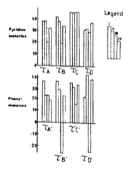

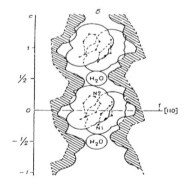

in determining the overall molecular shape of the host complex molecule.

If, instead of a methyl group, a substituent is introduced which adds its own conformational degree(s) of freedom, the problem becomes even more complex (Fig. 6)

The structures of the above type containing an ethyl [M. H. Moore et al., J. Chem. Soc., Dalton Trans. 1987, 2125] or a vinyl substituent [J. Lipkowski et al., J. Incl. Phenom. 2(1984), 317] have been found in inclusion-type compounds containing a different organic guest. It seems that the conformation of the end group is much more flexible than that of the coordinated pyridine rings. Even with bulky phenyl groups (Fig. 7), the conformational angles may vary from about -30 to 50 degrees (Fig. 8).

Fig. 8. Conformational angles of four pyridine and phenyl moieties in the complex molecule shown in Fig. 7.

Fig. 9. The structure of the channel-type inclusion compound - $Cd(NCS)_2$(4-Methylpyridine)$_4$ $\times 0.67$(4-Methylpyridine) $\times 0.33(H_2O)$

III. The crystalline structure of the host

The term 'Werner clathrates' comprises a large variety of crystal structures in the literature. There are channel-type inclusion compounds like $Cd(NCS)_2(4\text{-Methylpyridine})_4 \times 0.67(4\text{-Methylpyridine}) \times 0.33(H_2O)$ [N.V. Pervukhina et al., Zh. Strukt. Khim. 26(1985), 120] (Fig. 9) layered inclusion compounds like $Cu(NCS)_2(4\text{-Methylpyridine})_4 \times 2(4\text{-Methylpyridine})$ [J. Lipkowski et al., J. Incl. Phenom., submitted] (Fig. 10), and 'organic zeolite'

Fig. 10. Layered guest/host packing in $Cu(NCS)_2(4\text{-Methylpyridine})_4 \times 2(4\text{-Methylpyridine})$

Fig. 11. Packing of host $Ni(NCS)_2(4\text{-Methylpyridine})_4$ and guest benzene molecules (dotted) in 'organic zeolite' structure viewed along b axis.

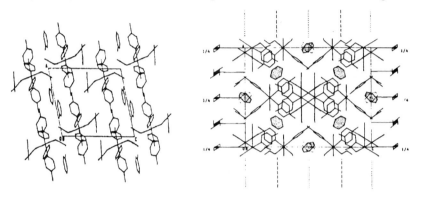

inclusion compounds in which clathrate cages form a three-dimensional network of interconnected cavities [J. Lipkowski et al., J. Mol. Struc. 75(1981), 101] (Fig. 11).

Table 2 summarizes structure data available for just one host complex which will also be used in part IV as a representative example.

IV. Guest-lattice interactions

In view of the very large variety of structures of Werner clathrates, discussion in this part will be limited to the example of inclusion compounds formed by $Ni(NCS)_2(4\text{-Methylpyridine})_4$. This is a convenient model since the Ni(II) oxidation state and octahedral coordination of the Ni complex are rather stable, NCS anionic ligand yields structures less soluble in most organic solvents than e.g. halide anions and, as illustrated in Table 2, the complex shows great versatility in clathrate formation. The host complex may form inclusion compounds not only through crystallization from a solution containing a suitable guest component but also by direct uptake of liquid or gaseous guest (Fig. 12)

Fig. 12. X-ray powder diffraction patterns of $Ni(NCS)_2(4\text{-Methylpyridine})_4$ in: non-clathrated 'alpha' form, 'beta' clathrate, and the product of reaction of solid 'alpha' substrate with liquid p-xylene at 25°C.

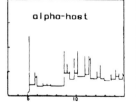

alpha-host

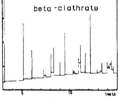

beta-clathrate

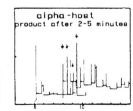

alpha-host
product after 2-5 minutes

Table 2.

Host: Ni(NCS)$_2$(4-Methylpyridine)$_4$

Guest [a)	Stoichiometry [b)		Crystal data		
Benzene	1:1	I4$_1$/a	a=16.98;	c=23.03	
4-Methylpyridine					
Toluene					
m- and p-xylene					
Methanol	2:1	"	a=16.99;	c=22.29	
ORGANIC ZEOLITE					
Naphthalene	2:1	C2/c	a=16.27;	b=16.46;	c=31.93
			β=89.3°		
1-Methylnaphthalene	2:1	P2$_1$/c	a=11.53;	b=11.89;	c=32.85
1-Bromonaphthalene			β=94.3°		
Azulene					
o-Xylene					
2-Methylnaphthalene	2:1	P$\bar{1}$	a=11.31;	b=9.58;	c=11.66
2-Bromonaphthalene			α=111.5;	β=82.0;	γ=108.7°
p-Terphenyl	1:1				
LAYERED CLATHRATE STRUCTURES					
Bromobenzene	2:1	Pnma	a=16.66;	b=15.69;	c=15.83
m- and o-Bromonitrobenzene	2/3:1	R3(?)	a=55.26;	c=11.08	
CLATHRATE STRUCTURES ?					

a) Alternative guests of isostructural series are listed in italic
b) given as the guest/host molar ratio

It may be seen that, after a few minutes of contact with xylene (guest), the solid complex is transformed into a clathrate structure. Since a similar effect occurs with gaseous xylene, it has been assumed that the clathrate formation process is (in this case) a structure transformation in the solid phase with simultaneous absorption of the guest. The same process studied by other techniques (dynamic calorimetry) appears as somewhat more complicated.

Fig. 13. The thermokinetic course of clathration of xylenes by Ni(NCS)$_2$(4-Methylpyridine)$_4$.

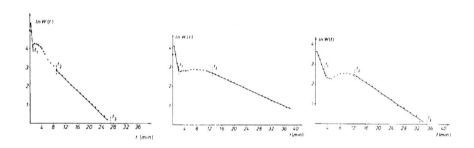

Fig. 13 shows the kinetics of the process given as heat evolved (ln W(t)). These results have been interpreted [J. Lipkowski et al., Thermochim. Acta,] as corresponding to a two-step processes: within the first step, lasting a few minutes, host lattice reconstruction takes place together with absorption of some guest. Some absorption sites in the clathrate are left empty at this stage. Then, within the second kinetic step absorption of the guest takes place up to maximum site occupancy. The above results have been quoted here in order to stress the non-stochio- metric nature of the inclusion compounds known as Werner clathrates. If the clathrate structure contains open cages, like the organic zeolite structure illustrated above, it may reversibly desorb-absorb different guest species. In the thermodynamic sense, the range of compositions expressed as the guest/host molar ratio (y) should be within y(min), which corresponds to the minimum content of the guest necessary for the clathrate system to be stable, and y = 1, i.e. full occupancy of all cavities. The y(min) value for the organic zeolite structure has been estimated as approx. 0.5. Moreover, the structure displays significant 'extrathermodynamic' stability. Volatile guest compounds (e.g. methanol or benzene) may be fully desorbed from the clathrate and the resulting 'empty' clathrate does not transform into the stable non-clathrated phase during several weeks. This allows for studying the complete range of absorption isotherms and the preparation of clathrates having noble gases, oxygen or nitrogen as the guest.

To learn the nature of the guest-lattice interactions let us refer to some thermochemical data and assume the simple criterium for distinguishing between 'specific' or 'non-specific' absorption used in zeolite chemistry [R.M. Barrer, "Zeolites and Clay Minerals as Sorbents and Molecular Sieves", Acad. Press, London, 1978] that the ratio of heat of absorption to the heat of liquefaction of the guest should be close to 2.0 if absorption is 'non-specific'. In Fig. 14 the straight line represents this criterion: below the line are the points representing absorption which is non-specific (the more distant the point from the line the weaker absorption) and the points above the line for specific absorption. Evidently the guest-lattice interactions are non-specific. Moreover, whereas in the sequence benzene - toluene - p-xylene (or ethylbenzene) there is a straight correlation of the heats of absorption, in the case of isomeric xylenes there are large differences which may be ascribed to steric interac-

Fig. 14. Graph of enthalpy of sorption versus enthalpy of evaporation of different guest compounds at 25°C. Host: Ni(NCS)$_2$(4-Methylpyridine)$_4$.

tions (negative, in the sense that m- and o-xylene are less favourably absorbed than more 'linear' p-xylene and ethylbenzene molecules) (Fig. 14).

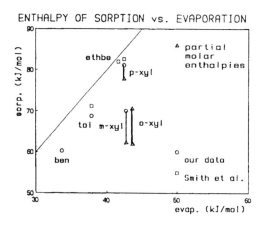

The range of absorption heats of each particular xylene isomer represent the range of experimentally determined dependence of partial molar heats of absorption on the composition of the clathrate phase. These quantities depend on the content of that particular guest component (Table 3). The interpretation of this experimental result, from the point of view of the clathrate structure, is that the clathrate host structure 'adapts' the shape and size of its cavities so as to meet steric requirements of the absorbed guest.

Table 3

CLATHRATION OF MIXTURES OF GUEST COMPOUNDS

$$\Delta H_{clath} = \sum_i (H_{clath})_i y_i$$

$(H_{clath})_i$ - partial molar enthalpy of clathration of the i-th guest

y_i - molar fraction of the i-th guest in the enclathrated mixture

Three-component mixture of o-, m- and p-xylene:

$y_o + y_m + y_p = 1$

The best-fit to experimental, calorimetric data given below:

$-(H_{clath})_o = 10.07 + 9.07 y_o$
$-(H_{clath})_m = 13.38 + 5.64 y_m$
$-(H_{clath})_p = 25.12 + 6.27 y_p$

Illustration of the results - two-component mixture of m- and p=xylene:

$-\Delta H_{clath} = 13.38 y_m + 5.64 y^2_m + 25.12 y_p + 1.50 y^2_p = 11.91 y^2_p + 0.46 y_p + 19.02$

Thus, if (say) p-xylene is absorbed as admixture to m-xylene (cavities adapted to m-xylene molecules), then its heat of absorption is less than when the main guest component is p-xylene. Although this experimental rule holds for every guest studied, the adaptation of the host structure shows its limitations since the heat of absorption of m- or o-xylene cannot attain the value found for p-xylene. Summarizing the above considerations it may be concluded:
(i)   clathration of a given guest component leads to formation of one of several possible structures of the host lattice - the guest 'selects' the most suitable crystal structure of the host;
(ii) the resulted structure shows significant flexibility - it may adapt its sorption centers within a range of size and shape;
(iii) the guest-lattice interactions are 'non-specific' - this may be referred to as a three-dimensional physical adsorption rather than a chemisorption process.
The problem of guest-lattice interactions in Werner clathrates is not limited to the questions discussed above. Thus, little is known yet about the dynamic structure of the compounds, on possible oxidation/reduction interactions between guest and host lattice, etc. Dynamic lattice properties may be of great importance when considering clathration selectivity towards isotopomers. It has been found that clathration by the host discussed here is selective enough in order to separate quantitatively (by liquid chromatography using the clathrate as stationary phase) some deuterated aromatic hydrocarbons from respective protiated species [N.O. Smith, in J.L. Atwood, J.E.D. Davies and D.D. MacNicol (Eds.), "Inclusion Compounds", vol. 3, ch. 8, Acad. Press, London, 1984]. From the data in Table 4 [J. Lipkowski et al., J. Incl. Phenom. and Mol. Recognition in Chem. 7(1989), 511] it may be seen that selectivity of clathration may be enthalpy or entropy driven, thus suggesting the importance of dynamic factors.

Table 4

Gibbs free energy ($\Delta G°$), enthalpy ($\Delta H°$) and entropy ($\Delta S°$) of guest exchange processes, and clathration enthalpy (ΔH_{clath}) of second guest in guest pairs.

	Guest pair	$\Delta G°$	$\Delta H°$	$\Delta S°$	$-\Delta H_{clath}$
1.	p-xylene/p-dibromobenzene	-6960	2900	33.0	28.5
2.	p-xylene/ethylbenzene	3220	-800	-13.5	32.2
3.	p-xylene/toluene	4260	7700	11.5	23.7
4.	p-xylene/4-Methylpyridine	5200	11300	20.5	20.1
5.	p-xylene/benzene	6350	11300	16.6	20.1
6.	p-xylene/o-xylene	6940	12200	17.7	19.2
7.	p-xylene/m-xylene	7930	12400	15.0	19.0

The problem of possible redox guest-host interactions in the clathrate lattices still awaits investigation, although there is some indication about such a possibility [J. Lipkowski, J. Incl. Phenom. and Mol. Recognition in Chem., submitted] in clathrate structures containing easily oxidable $Fe(NCS)_2$(4-Methylpyridine)$_4$ host. One may conclude, that this fascinating problem is still open, although some fundamental characteristics have been established.

CONCLUSIONS

Significant progress in our understanding of the physico-chemical nature of the compounds referred to as 'Werner clathrates' came from three main approaches:
~ 1950 - W.S. Schaeffer et al.:

   Clathrates - guest molecules isolated in cavities of the host crystal lattice;
~ 1960 - P. de Radzitzky et al.:

   Charge-transfer guest-host complexes - stacks of alternate guest and host molecules; and
1969 - S.A. Allison and R.M. Barrer:

   Organic zeolites - cavities in the host lattice interconnected to form a three-dimensional network; cavities are hydrophobic; host structure is expandable.

From the above discussion it seems clear that the last of these three ideas most closely corresponds to the nature of Werner clathrates or, rather to the so-called 'beta-form' with a zeolite-like system of cavities. More generally, the Werner clathrates have the character of interstitial solid solutions of the guest in a 'polymorph structure' of the host, by which one means a structural modification of the host complex which is not stable without any guest (although it may be metastable, as the organic zeolite structure described above, which may be prepared in its pure state, i.e. without any guest present). Other recent ideas appear to be consistent with the above, e.g. 'organic clays' [J.L. Atwood]. The author believes this proves that 'interstitial solid solution', far from being an empty class of compounds in organic crystal chemistry, actually includes clathrates, organic zeolites and organic clays as sub-classes.

4

Modelling steroid hormone receptor interactions on the basis of crystallographic and biochemical studies

W. L. Duax and J. F. Griffin

1. INTRODUCTION

Steroid hormones are vital to numerous physiological processes including cell growth, sexual development, maintenance of salt balance and sugar metabolism. Many of these activities are known to be contingent upon the binding of steroids to specific cytosolic protein receptors and the subsequent interaction of the steroid receptor complex with chromatin (Jensen *et al.*, 1962).

Steroids that differ from one another in subtle ways exhibit significant differences in their affinity for different receptors and associated biological activities. A careful examination of the molecular structures and three-dimensional shapes of the hormones, antihormones, chemicals and drugs that compete for a common binding site on a specific receptor can provide information on binding and the stereochemical requirements for the structural features that influence the extent and nature of the hormonal response.

Crystallographic data on over 400 steroids (Duax *et al.*: 1975, Griffin *et al.*, 1984) provide information concerning preferred conformations, relative stabilities and substituent influence on the interactive potential of steroid hormones.

On the basis of a comparison of the molecular structures, receptor binding and activities of several major classes of hormonal steroids including estrogens and progestins, an A-ring binding/D-ring active model has been proposed to mediate the activity of these hormones. In contrast, binding and activity data on androgens, suggest that in this case a D ring binding/A-ring active model may be more appropriate.

2. **ESTROGEN AGONISTS AND ANTAGONISTS**

The vast majority of structures with high affinity for the estrogen receptor contain a phenolic ring that appears to play a major role in initiating receptor binding (Duax *et al.*, 1980). With few exceptions steroids that compete most effectively for the estrogen receptor do not compete well for the other steroid hormone receptors. This high degree of selectivity of potent estrogens for the estrogen receptor distinguishes estrogens from other steroids (Raynaud *et al.*, 1979). Such simple molecules as tetrahydronaphthol and p-sec-anyl phenol have been used to inhibit the binding of estradiol or displace estradiol from its binding site (Mueller *et al.*, 1978) further illustrating the importance of the phenolic ring and demonstrating that the intact steroid or an intact D ring is not critical to binding. The fact that the removal of the 3-hydroxyl group from estradiol reduces the binding affinity, to less than 2%, illustrates the importance of the hydroxyl to A-ring binding (Chernayaev *et al.*, 1975). The removal of the 17-hydroxyl also reduces binding, but to a lesser extent.

When the phenol rings of a sample of the compounds that compete for binding to the estrogen receptor are superimposed, significant differences in the D ring region of the molecules are observed (Fig. 1). If there is a close association between estrogens and the receptor, it would appear to be limited to the A and B rings. The receptor is either flexible in the D ring region or insensitive to it (Duax *et al.*, 1978a; Duax *et al.*, 1980).

The synthetic estrogen diethylstilbestrol has two phenolic rings capable of imitating the A-ring of estradiol in initiating receptor binding. The relative orientation of the hydroxyl groups of the two phenyl rings and the distance between them is fixed by the chemical components of DES so that the overall shape of the molecule closely resembles that

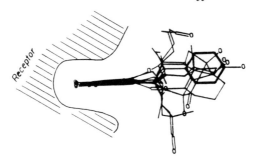

Figure 1. Superposition of the phenol rings of six compounds in Figure 1 that bind to the estrogen receptor, suggests that variability in D-ring orientation is compatible with receptor binding and some degree of activity.

of estradiol (Fig. 2a). This similarity in shape and positioning of hydrophobic groups has suggested that the location of the D-ring hydroxyl group relative to the phenolic A-ring may be a critical factor in determining estrogen activity (Duax *et al.* 1980).

Analogues and metabolites of DES are of interest due to uncertainty concerning the form responsible for the carcinogenic properties of DES and as additional probes of the structure-activity relationships of estrogens. It is evident that the precise location of steric bulk and functional groups relative to the phenolic ring will govern binding and activity. Pseudo-DES differs from DES in the location of the double bond and exists as the Z and E isomers, ZPD and EPD. While both forms bind to the estrogen receptor, only the Z form has appreciable activity. X-Ray structure determinations of the two compounds, revealed that both EPD and ZPD have bent conformations (Fig. 2b and c), completely unlike the conformations of estradiol and DES. When the crystallographically observed conformations of EPD and ZPD were subjected to energy minimization the structures retained their overall conformations with only minor changes in individual torsion angles that averaged less than 4° (Duax *et al.*, 1985).

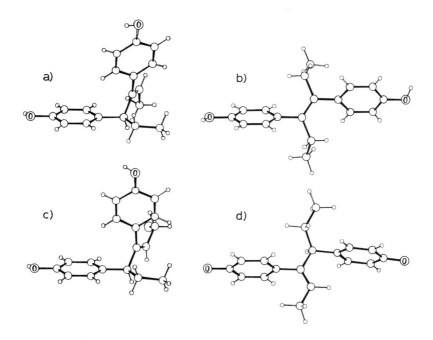

Figure 2. Crystallographically observed conformations of (a) EPD, (b) DES, and (c) ZPD and the local energy minimization of ZPD (d) that resembles DES.

The consistency between the crystallographic results and the energy calculations suggests that the bent form is the minimum energy conformation of ZPD and EPD. However, previous studies indicate that this conformation is unlikely to be compatible with any significant degree of estrogenic biological activity (Duax *et al.*, 1980).

Potential energy calculations on extended conformers of ZPD and EPD revealed interesting differences. As a result of extremely close contacts (less than 2.2Å) between the hydrogens on a methyl group and carbons of the α-ring in EPD, the extended conformation is energetically unfavorable and the

molecule refined to a bent conformation resembling the
crystallographically observed structure in overall shape.   In
contrast to this, the extended conformation of the more active
Z isomer does not incorporate intolerable non-bonding
interactions and refinement indicates that a local stable
minimization energy conformation exists which resembles the
DES conformation (Fig. 2d), satisfactorily accounting for the
observed activity.   The relative potential energies of the
crystallographically observed conformer and the extended
conformer of ZPD (Duax *et al.*, 1985) suggest that only a small
fraction of a population of ZPD molecules, perhaps 10%, are in
an active conformation.   This could account for the fact that
ZPD is only one tenth as active as DES, which is constrained
to have the active, extended conformation at all times (Duax
et al., 1980).

     Estrogen antagonists such as *trans*-tamoxifen compete for
binding to estrogen receptors and elicit little or none of the
characteristic hormonal response.   Although at least two types
of binding sites are present on the estrogen receptor (Clark
et al., 1980), it is clear that the clinical utility of the
estrogen antagonists is a result of their competition for the
traditional Type I estrogen binding site (Clark *et al.*, 1973).

     The 4-hydroxy metabolite of *trans*-tamoxifen has been
shown to be the potent competitor for the estrogen receptor
that is probably responsible for the antagonist properties
useful in breast cancer therapy (Borgna *et al.*, 1982).   This
observation has led most investigators (Duax *et al.*, 1981a;
Durani *et al.*, 1979; Pons *et al.*, 1984) to conclude that, at
least in the case of the 4-hydroxy derivative of tamoxifen, it
is the hydroxylated ring that mimics the estradiol A-ring in
receptor interaction.   The molecular fit achieved by
overlapping the α-ring of tamoxifen and the A-ring of
estradiol is illustrated in Fig. 3a.   The antiestrogen has the

phenolic ring required for high affinity binding to the receptor, but lacks a hydrogen bonding group corresponding to the O(17) hydroxyl of estradiol that appears to be required for activity. The antiestrogen also possesses a bulky substituent extending nearly perpendicular to the steroid skeleton that may interfere with a conformational change in the receptor that is postulated to be essential to subsequent macromolecular interaction and activity (Duax *et al.*, 1981a).

3. PROGESTIN AGONISTS AND ANTAGONISTS

Examination of the chemical structures of steroids having high affinity for the progesterone receptor indicated that extensive structural variation is compatible with high affinity binding (Duax *et al.*, 1978b). The only structural features common to almost all compounds with high affinity for the progesterone receptor is the steroid ring system and 4-en-3-one composition.

Many steroids that have the 4-en-3-one composition have little or no affinity for the progesterone receptor (e.g. testosterone). Since a 4-en-3-one A ring appears to be required, but not sufficient for high-affinity binding to the progesterone receptor, we examined the conformations of the 4-en-3-one A rings of the compounds with highest affinity in search of some unusual electronic, geometric, or stereochemical feature that might explain their enhanced binding. Many of the modified steroids having highest affinity for the progesterone receptor were observed to have a characteristic A-ring conformation that differed from the conformation found in most naturally occurring steroids. Rather than the 1α,2β-half chair conformation commonly found in naturally occurring 4-en-e-one steroids, a disproportionate number of the most potent synthetic progestins were found to have an inverted 1β,2α-half chair conformation. Consequently,

we proposed that the inverted 1β,2α-conformation of the A ring
found in the potent progestin R5020 and medroxy progesterone
acetate is ideally suited to high affinity receptor binding
(Duax *et al.*, 1978b)

A series of 11β-substituted 19-norsteroids have proven to
be potent antiprogestins. One of the most effective of this
series is 11β-(4-dimethylaminophenyl)-17β-hydroxy-17α-(prop-1-
ynyl)-estra-4,9-diene-3-one (Teutsch, 1984). X-Ray analyses
of other members of the series having an 11β-phenyl
substituent have shown that these potent antagonists have the
inverted 1β,2α-half-chair conformation. The
crystallographically observed conformations of the potent
progestin 17,21-dimethyl-19-nor-4,9-pregnadiene-3,20-dione,
R5020 (Busetta *et al.*, 1974) and the antiprogestin, 11β-p-
methoxyphenyl-17β-hydroxy-19-nor-17α-pregna-4,9-dien-20-yn-3-
one, RU25056 (Surcouf, 1982) are compared in Figure 3b. The
fact that the most potent progestin agonist and antagonist
have in common an unusual conformation of the A ring further
supports the A-ring binding/D-ring acting model for steroid
action (Duax *et al.*, 1984).

The similarity in steric and electronic properties of the
A rings of potent progestins and antiprogestins may account
for their competition for a common binding site on the
receptor. Although very bulky 11β-phenyl substituents can be
tolerated by the receptor and may in fact contribute to
enhanced binding, some appear to interfere with the
conformational change in the receptor that is essential for
enhanced activity. The blocking of conformational change
would be analogous to the molecular basis for the
antiestrogenic behavior of tamoxifen.

4.   ANDROGEN AGONISTS AND ANTAGONISTS
The binding affinity of hundreds of steroids for the

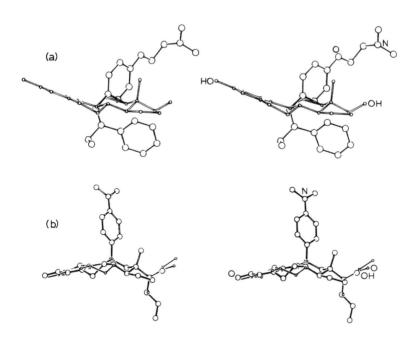

Figure 3. (a) Stereo superposition of an estrogen agonist, estradiol (small circles and antagonist, 4-hydroxy-tamoxifen (large circle. (b) Stereo superposition of a progestin agonist, (R5020) (small circles) and the potent antagonist RU25056 (large circles).

androgen receptor in rat prostate (Cunningham *et al.,* 1979; Kirchhoff *et al.,* 1979) suggests that a 17β-hydroxy and a 3-oxo group are essential for high affinity binding. However, there have been reports of A-norsteroids exhibiting appreciable androgenic activity (Zanati *et al.,* 1973). Most evidence suggests that if any portion of the androgen is critical to receptor binding it is the 17β-hydroxy substituted D ring. Steroidal antiandrogens that are of sufficient potency to be used clinically also possess a 17β-hydroxy substituted D-ring (Hamada *et al.,* 1963; Nakayama *et al.,*

1979; Mangan *et al.*, 1972). The 17β-hydroxyl group is usually flanked by an 18-methyl group and a hydrophobic substituent in the 17α or 16β position. These groups may be directly involved in receptor interaction or have some influence on the orientation and strength of the interaction of the 17β-hydroxy group with the receptor.

Several nonsteroidal compounds that compete for the androgen binding site and alter hormonal response provide additional clues to the requirements for receptor binding and activity. Liao has shown that polyunsaturated tricyclic compounds such as 9,10-dihydrophenanthrene (Liao *et al.*, 1984) compete for the androgen binding site. He has proposed that the androgen binding site may act as a cavitand with high affinity for small, fairly planar molecules that lack the structural features needed to promote the requisite conformational change in the receptor.

One of the most active antiandrogens is the nonsteroidal compound flutamide (Wakeling *et al.*, 1981). The X-ray crystal structure determination of hydroxy-flutamide revealed that is an extended planar molecule in which sixteen of the twenty nonhydrogen atoms are essentially coplanar. Model fitting reveals that flutamide combines the planar character of the Liao antiandrogen with the 17-hydroxy group flanked by hydrophobic methyl groups characteristic of the 17-hydroxyl environment typical of the most potent androgens.

Since the outline of the flutamide molecule appears to fall within that of the androgen agonists (Fig. 4). it would appear to function in a passive manner, failing to have some feature essential to induction or stabilization of receptor transformation or requisite macromolecular interaction that follows hormone binding. In contrast, the antagonist action of tamoxifen and RU25056 could be active, a result of the fact that the very bulky hydrophobic substituent, projected perpendicular to the general plane of the steroid, **blocks a** necessary conformational change in the receptor.

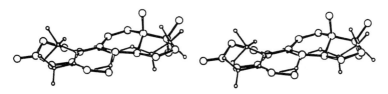

Figure 4.   A reasonably good match between the steric and electronic features of the potent androgen agonist 17β–hydroxy–17α–methyl–4,9,11–estratrien–3–one (R1881) and the potent androgen antagonist hydroxyflutamide (HYF) is achieved when the phenyl ring of HYF is aligned with the B–ring of R1881 and the C(9) atom of HYF is fitted to C(17). This aligns the flat HYF molecule with the planar 3,9,11–triene configuration of R1881. The positioning of the hydroxyl substituent on C(9) of HYF flanked by two methyl groups bears a resemblance to the position of the 17β–hydroxyl between the 18– and 17α–methyl groups in R1881.

5. MODELING RECEPTOR STRUCTURE AND INTERACTION

The steroid receptors are a family of soluble proteins having extensive sequence homologies. The receptors range in size from 60 kD (Vitamin D, McDonnell *et al.*. 1987) to 107 kD (mineralocorticoid, Arriza et al., 1987). Analysis of the sequences of the receptors and preparation of mutants have led to the characterization of several domains in the steroid receptor, the DNA binding domain, the steroid binding domain and an immunodominant domain that appears to have a role in transcriptional activation.

The steroid binding region is at the C–terminal portion of the receptor. The degree of sequence homology observed in the steroid binding region of the receptor parallels the competition for binding to different sites on different receptors by the steroid hormones. Functional studies of the

glucocorticoid receptor suggest that hormone binding unmasks
the DNA binding site to permit interaction of the receptor
with DNA and activation of transcription (Godowski *et*
al..1987; Hollenberg *et al.*. 1987).

The technique of affinity labeling and protein digestion
have been used to further define the binding site within the
C-terminal domain. Ligands used to label the glucocorticoid
receptor have included ^{3}H-triamcinolone acetonide (TA) and H_3-
dexamethasone mesylate (DM) (Carlested-Duke *et al.*. 1988;
Simons, 1987). UV-irradiation of triamcinolone acetonide
produces free radicals of the A ring of the steroid (Benisek,
1977) that are assumed to interact with nearby amino acid
residues. The α-mesylate group will interact specifically
with cysteine residues. Using such probes it has been
possible to demonstrate their covalent attachment to Cys 656,
Met 622 and Cys 754. The data suggest that attachment to Cys
656 is through the corticoid side chain and that Met 622 and
Cys 754 are bound to the A-ring.

The data also suggest that the steroid binding domain is
folded to form a hydrophobic pocket with residue 622 and 752
in close proximity to one another and about 12 to 16 Å from
residue 656.

The cocrystallization of steroids with amino acids and
nucleic   acids could provide evidence of stereospecific
interaction and models for steroid macromolecular binding.
Although spectral evidence for such interaction has often been
obtained, cocrystallization has rarely been achieved.
Progesterone is unusual in that it cocrystallizes readily with
hydrogen bond donors including indole, and resorcinol. The
progesterone complex with indole is of particular interest
because ultraviolet spectra indicate an interaction between
the 4-en-3-one chromophore and tryptophan residues when
progesterone is bound to progesterone binding globulin
(Stroupe *et al.*. 1975). In the crystal structure, indole

molecules are hydrogen bonded to the carbonyl oxygens O(3) and O(20) as illustrated in Figure 5. The indole molecules do not stack over each one or over the 4-en-3-one chromophore. The progesterone resorcinol complex exhibits analogous interaction.

A tryptophan residue (TRP 618) near the site of attachment of triamcinolone acetonide (Met 622) in the rat liver glucocorticoid receptor may well participate in steroid binding via an interaction comparable to that seen in the progesterone indole complex. A homologously located tryptophan residue is present in glucocorticoid, mineralocorticoid and progesterone receptors (Evans *et al.*. 1988). Substitution of a nonpolar residue for Trp 618 would test the hypothesis that such a tryptophan A ring interaction makes a significant contribution to receptor binding.

Quantities of pure stable proteins sufficient to allow the pursuit of single crystal X-ray analysis of any of these receptors are not yet available. At the present time, the trp repressor may provide a most useful model for steroid receptor-DNA interaction. Two crystal forms of the trp repressor and one of the inactive unliganded aporepressor have been refined to atomic resolution (Lawson *et al.*. 1988). The dimeric structure is made up of a central core flanked by two flexible "reading heads". The binding of tryptophan molecules to sites between the core and reading heads force the heads apart so that they can penetrate successive major grooves of B-DNA. The crystal structure of a complex between the trp repressor and its DNA operator region revealed the direct participation of the tryptophan molecule in DNA intraction, forming a 2.9Å hydrogen bond to a phosphate oxygen (Sigler *et al.*. 1988).

The binding of desamino analogues of L-tryptophan such as indole propionate causes formation of an inactive

pseudorepressor. X-Ray analysis of a complex of the trp
repressor with indole propionate (IPA) revealed that the IPA
binds in the same position as the indole but is "flipped
over". As a result, the carboxyl group of IPA is oriented
towards the DNA binding surface of the protein and is in a
position where it sterically and electrostatically repels the
phosphate backbone of both operator and non-operator DNA
(Lawson *et al*., 1988). It is conceivable that when the
receptor-steroid complex interacts with DNA, the steroid D-
ring may be sufficiently exposed to contact the DNA directly.
A possible model for such an interaction is provided by the
crystal complex of deoxycorticosterone and adenine (Weeks, *et
al*., 1975); the carbonyl and hydroxyl substituents on the
corticoid D-ring form hydrogen bonds to the two nitrogens of
adenine that would normally be involved in Watson-Crick base
pairing. Such contacts might be critically involved, either
in DNA sequence recognition or in the activation of
transcription by the steroid-receptor complex.

6. SUMMARY

Examination of the structures of compounds having high
affinity for estrogen and progestin receptors strongly
suggests that receptor binding is primarily the result of a
tight association between the receptor and the steroidal A-
ring. High affinity binding to the estrogen receptor appears
to be dependent upon the presence of a phenolic ring in the
substrate. An inverted $1\beta,2\alpha$ conformation of the 4-en-3-one A
ring appears to be most conducive to high-affinity binding to
the progesterone receptor. Antagonists that compete for
estrogen and progesterone receptor sites may be expected to
have the A-ring composition and conformation necessary for
receptor binding but lack the D-ring conformational features
and functional groups that induce or stabilize subsequent

receptor functions. Antagonists might also be compounds with A-ring conformations appropriate for binding but other structural features that interfere with subsequent receptor functions essential to activity. In contrast, androgen receptor binding data and molecular modelling studies suggest that in this case a D-ring binding/A-ring active model is more appropriate. Androgen receptor binding appears to be most dependent upon the presence of a 17β-hydroxy substituent and enhanced by the presence of a planar system of conjugated bonds in the B- and C-ring region. Compounds having the features conducive to high affinity binding to the receptor but lacking the presence of a suitable functional group at the A-ring end of the steroid would act as antagonists. The possible means by which the D-ring of estrogen, progestins and corticoids might control activity include: (1) inducing or stabilizing an essential conformational state in the receptor (allostery); (2) influencing the aggregation state of the receptor; or (3) participating in a direct interaction with DNA or chromatin.

ACKNOWLEDGEMENTS: Research supported in part by NIAMDD Grant No. DK26546. The authors wish to express their appreciation to J. Gallmeyer for assistance in the organization and preparation of this manuscript.

REFERENCES

ARRIZA, J. L., WEINBERGER, C., CERELLI, G., GLASER, T. M., HANDELIN, B. L., HOUSMAN, D. E. and EVANS, R. M. (1987). *Science* **237**, 268-275.

BENISEK, W. F. (1977). *Methods Enzymol.* **46**, 469-479.

BORGNA, J. L., COEZY, E. and ROCHEFORT, H. (1982). *Biochem. Pharmacol.* **31**, 3187-3191.

BUSETTA. B., COMBERTON, G., COURSEILLE, C. and HOSPITAL, M. (1974). *Acta Cryst.* **B30**, 2757-2759.

CARLSTEDT-DUKE, J., STROMSTEDT, P. -E., PERSSON, B., CEDERLUND, E., GUSTAFSSON, J. Å. and JORNVALL, H. (1988). *J. Biol. Chem.* **263**, 6842-6846.

CHERNAYAEV, G. A., BARKOVA, T. I., EGOROVA, V. V., SKOLOVA, N. A. and ROZEN, V. B. (1975). *J. Steroid Biochem.* **6**, 1483-1488.

CLARK, J. H., ANDERSON, J. and PECK, E. J. (1973). *Steroids* **22**, 707-718.

CLARK, J. H., MARKAVERICH, B., UPCHURCH, S., ERIKSSON, H., HARDIN, J. W. and PECK, E. J. (1980). *Recent Prog. Horm. Res.* **36**, 89-134.

DUAX, W. L. and NORTON, D. A. (1975). "Atlas of Steroid Structure", Vol. 1 Plenum Press, New York.

DUAX, W. L. and WEEKS, C. M. (1980). "Estrogens in the Environment", pp. 11-31. Elsevier, New York

DUAX, W. L., CODY, V., GRIFFIN, J. F., ROHRER, D. C. and WEEKS, C. M. (1978a). *J. Toxicol. Environ. Health.* **4**, 205-227.

DUAX, W. L., CODY, V., GRIFFIN, J., HAZEL, J. and WEEKS, C. M., (1978b). *J. Steroid Biochem.* **9**, 901-907.

DUAX, W. L., GRIFFIN, J. F., ROHRER, D. C., SWENSON, D. C. and WEEKS, C. M. (1981a). "Recent Advances in Steroid Biochemistry" Vol. 15, pp. 41-47. Pergamon, New York.

DUAX, W. L. and GRIFFIN, J. F. (1984). "Adrenal Steroid Antagonism, pp. 15-41. Walter de Gruyter and Co., Berlin-New York.

DUAX, W. L., GRIFFIN, J. F., WEEKS, C. M. and KORACH, K. S. (1985). "Environmental Health Perspectives", Vol. 61, pp. 111-121. Elsevier, New York.

DURANI, S., AGARWAL, A. K., SAXENA, R., SETTY, B. S., GUPTA, R. C., KOLE, P. L., RAY, S. and ANAND, N. J. (1979). *Steroid Biochem.* **11**, 67-77.

EVANS, R. M. and HOLLENBERG, S. M. (1988). *Cell* **52**, 1-3.

GODOWSKI, P. J., RUSCONI, S., MIESFELD, R. and YAMANOTO, K. R. (1987). *Nature* **325**, 365.

GRIFFIN, J. F., DUAX, W. L. and WEEKS, C. M. (1984). "Atlas of Steroid Structure", Vol. 2. Plenum Press, New York.

HAMADA, H., NEUMANN, F. and JUNKMANN, K. (1963). *Acta Endocrinol. Copenhagen* **44**, 380-388.

HOLLENBERG, S. M., GIGUERE, V., SEGUI, P. and EVANS, R. M. (1987). *Cell* **49**, 39.

JENSEN, E. V. and JACOBSON, H.I. (1962). "Recent Progress in Hormone Research", pp. 387-414. Academic Press, New York.

LAWSON,, C. L. and SIGLER, P. B. (1988). *Nature* **33**, 869–871.

LAWSON, C. L., CHANG, R.-G., SCHEVITZ, R. W., OTWINOWSKI, A., JOACHIMIAK, A. and SIGLER, P. B. (1988). *Proteins: Structure Function and Genetics* **3**, 18–31.

LIAO, S., WITTE, D., SCHILLING, K. and CHANG, C. (1984). *J. Steroid Biochem.* **20**, 11–17.

MANGAN, F. R. and MAINWARING, W. I. P. (1972). *Steroids* **20**, 331–343.

MC DONNELL, D. P., MANGELSDORF, D. J., PIKE, J. W., HAUSSLER, M. R. and O'MALLEY, B. W. (1987). *Science* **235**, 1214–1217.

MUELLER G. and KIM. U.-H. (1978). *Endocrinology (Baltimore)* **102**, 1429–1435.

NAKAYAMA, R., MASUOKA, M., MASAKI, T. and SHIMAMOTO, K. (1979). *Acta Endocrinol. 92/Suppl.* **229**, 2–23.

PONS, M., MICHEL, F., CRASTES DE PAULET, A., GILBERT, J., MIQUEL, J.-F., PRECIGOUX, G., HOSPITAL, M., OJASOO, T. and RAYNAUD, J. P. (1984). *J. Steroid Biochem.* **20**, 137–145.

RAYNAUD, J. P., OJASOO, T., BOUTON, M. M. and PHILIBERT, D. (1979). "Drug Design, Vol. 8, pp. 169–214. Academic Press, New York.

SIGLER, ,P. B., OTWINOWSKI, Z., SCHEVITZ, R. W., ZHANG, R. -G., LAWSON, C. L., MARMORSTEIN, R. Q. JOACHIMIAK, A. and LUIS, B. (1988). *Am. Crystallogr. Assoc. Annual Meeting* Abstract 08, p. 50.

SIMONS, S. S. (1987). *J. Biol. Chem.* **262**, 9669-9675.

STROUPE, S. D. and WESTPHAL, U. (1975). *Biochemistry* **14**, 3296-3300.

SURCOUF, E. (1982). Doctoral Thesis, University Pierre and Marie Curie. pp. 94-96, Paris.

TEUTSCH, G. (1984). "Adrenal Steroid Antagonism", pp. 43-75. Walter de Gruyter and Co., Berlin-New York.

WAKELING, A. E., FURR, B. J. A., GLEN, A. T. and HUGHES, L. R. (1981). *J. Steroid Biochem.* **15**, 355-359.

WEEKS, C. M. ROHRER, D. C. and DUAX, W. L. (1975). *Science* **190**, 1096-1097.

ZANATI, G. and WOLFF, M. E. (1973). *J. Med. Chem. 16. 90.*

5

Crystallographic studies of inter- and intra-molecular interactions

Tadeusz M. Krygowski

Inter- and intra-molecular interactions of chemical species are of primary importance in chemical
research. They influence chemical reactivity and are
associated with changes in the geometry of the chemical species involved. Hence, knowledge of the precise geometry of chemical systems is valuable as the
starting point for understanding their chemical, physical and biological properties (Hoffmann, 1983).
The most effective way to study molecular geometry is
by crystallography with X-ray and (to a lesser extent)
neutron diffraction on crystals. The thousands of
crystal and molecular structures solved and published
annually mean that thousands of geometries of chemical
species become available for interpretation. Two questions then arise:
   (i) is the molecular geometry determined in crys-
       tals comparable to that of the isolated che-
       mical species?
and then, if the lattice forces, i.e. intermolecular
interactions in crystals, affect the molecular geometry
   (ii) what kind of chemical information can be de-
        rived from these geometries?
   Before we answer these questions let us say a few
words about the geometry itself. The molecular shape
is most often described by the following geometrical
parameters: bond lengths, bond- or valence-angles,
torsion angles and dihedral angles. Deformations of
these parameters from some ideal or reference state
are associated with the values of force constants,
collected in Table 1 (Bernstein 1985). If a simple
harmonic oscillator model of energy of deformation is
applied, then for 1% deformation of the bond length
(elongating from 140.0 to 141.4 A) one needs 0.45
$kJ \cdot mol^{-1}$. On the other hand, if two carbon atoms ap-
roach to distance r = 340 pm (i.e. much closer than
their sum of Van der Waals radii, 177 x 2 = 354 pm)

Table 1. Force constants for some typical deformations in organic systems (Bernstein 1985)

Kind of deformation	$10^5 \cdot k$ in dyn cm^{-1}
Stretching of bonds: C - C C = C C = C C - C	4.5 9.6 15.7 7.6
in-plane deformation of CCC in benzene	0.7
out-of-plane of the ring in benzene	0.06

(Bondi 1964), the energy of repulsion calculated by
use of the Dashevsky potential (Kitaygorodzky 1987)
is E_{rep} = 0.27 kJ·mol^{-1}. Evidently, even if the calculations are very approximate, these two energies,
are of the same order of magnitude so that even the
hard structural parameters (such as bond lengths and
bond angles) may be affected by deformations due to
intermolecular interactions in the crystal (packing
forces). This is contradictory to the earlier asser-
tion by Kitaygorodzky (Kitaygorodzky 1973) who
held that hard structural parameter are not deformed
in crystals. This problem will be discussed in more
detail in the next section. It is, however, convenient to classify the geometrical parameters into
the hard and soft parameters. The hard parameters,
i.e. bond lengths and bond angles (endocyclic in
rigid cyclic systems, e.g. aromatics) are not easily
deformed by packing forces, unlike the soft ones
which are very flexible. All other parameters should
be considered as soft, including H-bridges and interatomic distances in EDA-complexes and related situa-
tions.
    While looking through the recent literature one
can easily find that soft structural parameters are
often applied to simulate the path of chemical reaction via the Bürgi and Dunitz (Dunitz 1979) principle
of structural correlation, as reviewed by Burgi and
Dunitz (1983). Sometimes they are used to build up
statistical models relating the changes in geometry to
the kind of data distribution and indirectly to the
energy scale of interactions involved in making deformations due to the crystal lattice (Murray-Rust 1982,
Jaskólski 1986). Because of their ease of deformation,

the soft parameters are not useful to study intramolecular interactions, e.g. substituent effects or
protonation effects, on the geometry of the system in
question. That is why problems of this kind are ex-
cluded in the present review.

<u>Application of the simple force field to study</u>

<u>the effect of packing forces on geometry of</u>

<u>chemical species</u>.

In order to estimate more quantitatively the
problem of the effect of packing forces on the hard
parameters let us consider the following results. The
crystal and molecular structure of sodium p-nitroben-
zoate monohydrate (hereafter its anion is abbreviated
as pNB⁻) was solved with high precision (Turowska-Tyrk,
Krygowski, Gdaniec, Häfelinger and Ritter, 1988)
(Fig. 1).

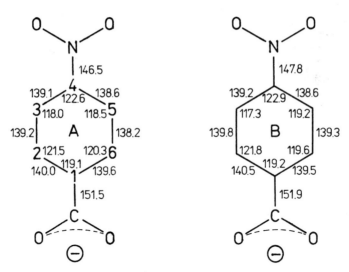

Fig. 1. Geometry (bond angles in degrees) of p-nitro-
        benzoate anion in its monohydrated salt with
        sodium (Turowska-Tyrk et al. 1988). Two inde-
        pendent anions in an asymmetric unit.

For an isolated pNB⁻ one could expect at least 2m symmetry. Hence breaking this symmetry, expressed by the
deformation parameter, DP,

$$DP = \varphi_i - \hat{C}_2\, \varphi_i = \varphi_i - \varphi_i' \qquad (1)$$

may be a quantitative measure of deformation due to
intermolecular interactions. Fig. 2 illustrates the
problem and Fig. 3 presents DP-values expressed in
esd units.

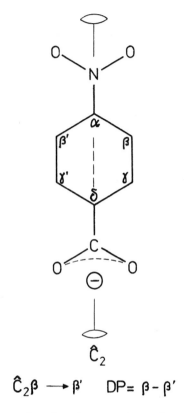

$$\hat{C}_2\beta \longrightarrow \beta' \qquad DP = \beta - \beta'$$

Fig. 2. Assignments of
    angles of p-nitroben-
zoate anion and $\hat{C}_2$ symmetry
operator used in calculation
    of deformation parameter,
DP, with the scheme of
using $\hat{C}_2$ to β -angles.

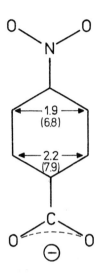

Fig. 3. Differences
    between β and β'
as well as γ and γ'
(i.e. DP-values for β
and γ) expressed in
degrees (esd in par-
entheses).

Clearly these deformations are statistically significant: for three of four $DP \gg 3\sigma$ (where σ was calcu-
lated for a difference $\varphi_i - \varphi_i'$ as $\sigma_{ii} = (\sigma_i^2 + \sigma_i^2)^{1/2}$. This result requires rationalization and it
was done (Krygowski and Turowska-Tyrk, 1987) within
the force field in which the energy of interatomic
repulsion (i.e. scalar quantity) was transformed into

the vector quantity, i.e. the force

$$\vec{F}_{ij} = - \text{grad } V_{ij} = - \frac{dV_{ij}}{d\vec{R}_{ij}} \tag{2}$$

where V_{ij} is the atom-atom potential of repulsive interaction between atoms i and j, whereas $\vec{R}_{ij}$ is the vector distance between those atoms. Thus, for the ith atom one can calculate the resultant force $\vec{F}_i$

$$\vec{F}_i = \sum_{j=1}^{n} \vec{F}_{ij} = - \sum_{j=1}^{n} \frac{dV}{d\vec{R}_{ij}} \tag{3}$$

for all close contacts for which $R_{ij} <$ sum of Van der Waals radii of atoms i and j. For these cases the force is repulsive (Fig. 4).

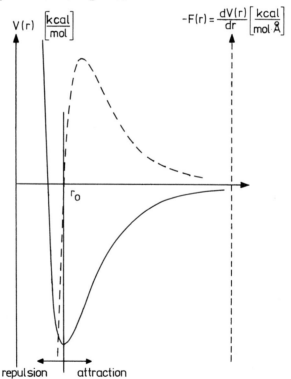

Fig. 4. Scheme of dependence of V(r) and −F(r) on the interatomic distance r.

These repulsive interactions are only local since for the crystal as a whole the resultant force of all interacting forces must be equal to zero. This, however, does not mean that in some parts of the crystal strong attractive forces may operate with strong repulsive ones in other parts. Of course, they must be balanced in the crystal as a whole.

This approach was used for the geometry of p-NB⁻ applying Kitaygorodzky (1973) parameters of the Lenard-Jones potential. Then, for every carbon atom of two independent pNP ions, the resultant forces F_i were calculated (Fig. 5) and compared with a nonadditivity parameter of an angle i at which the force was calculated.

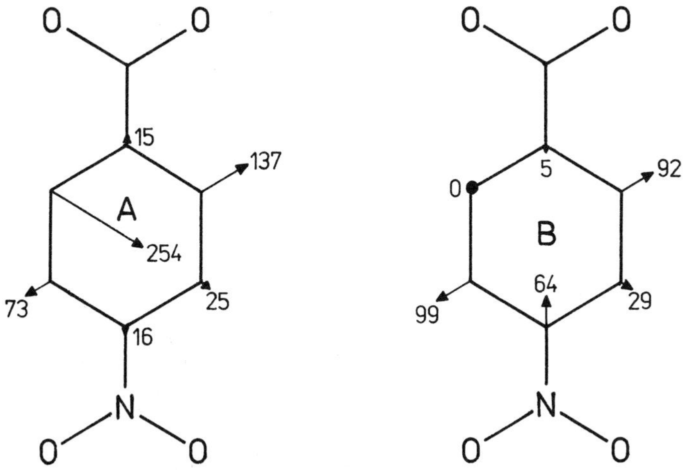

Fig. 5. Resultant forces $\vec{F}_i$ obtained by use of eq. (3) projected on dihedral of the endocyclic bond angle at i^{th} carbon atom in the ring.

Since both substituents, NO_2 and COO^-, are axially symmetrical and weakly interacting, the changes in endocyclic bond angles φ_i should follow the additivity scheme while substituent angular parameters $\Delta \varphi_i^o$ (Domenicano, Murray-Rust 1979) are used. Thus the quantity which measures a non-additivity effect, NP_i, should exhibit some trend of a linear (or at least monotonic) dependence on F_i. The non-additivity parameter NP_i is defined as

$$NP_i = \varphi_i(\text{measured}) - (120 + \Delta \varphi_i^o) \qquad (4)$$

The above two quantities are statistically independent; one, F_i, stems from interatomic distances of atoms in different ions, whereas the other, NP_i, was estimated from some reference set of data and positions of atoms in the frame of an ion. Thus, when plotting them one against the other, we should test the null hypothesis which says that these two sets of data are randomly distributed, i.e. that the correlation coefficient $R = 0$. The R-value obtained for this plot is 0.823 and, since the upper limit for $R = 0$ (for significance level $\alpha = 0.01$) is $R_{UL} = 0.719$ for $n = 11$, we have to reject the null hypothesis and accept the alternative one which reads:

NP_i-values depend on F_i values in a statistically significant way.

The physical meaning of this conclusion is that the greater is the repulsive force acting at atom j and towards the ring the greater becomes ℓ_j and hence the greater is the value of NP_j.

Taking into account the complexity of interactions and the considerable simplifications applied in calculating (2) and (3), it would seem that the result presented is encouraging for further research.

We conclude that, if we are going to study intramolecular interactions by analysis of molecular geometry determined from the crystalline state, we should be aware of the possibility of deformation due to intermolecular interactions. The weaker are the intramolecular interactions taken into account, the more thorough should be the procedure preventing the contamination of data by deformation due to the intermolecular interactions.

Let us add one more example to illustrate the situation still better. Domenicano, Vaciago and Coulson (1975) showed that the substituent effect on geometry of the ring of monosubstituted benzene derivatives follows the Walsh rule, i.e. an increase in electronegativity of the substituent should imply two structural effects: an increase of α-angle at the ipso carbon and a shortening of both CC-bond lengths (called a) linking the ipso position with both ortho positions. A rough dependence of α vs χ (electronegativity) was found but, curiously enough, not of a vs α. Variation of a-bond lengths for these cases was not large, probably close to the level of error, so that the a vs α plot failed. Replacement of the a-bond length by $b - a = \Delta$, that is, the difference between the central bond b and the a-bond, improved the situation (since a part of the error of estimation could be cancelled in subtraction)(Krygowski 1984). When the gas-phase electron diffraction and microwave geometries were taken into the $\triangle$ vs α plot, the correlation co-efficient r rose to 0.994; clearly the packing forces caused deformations which interfered too much for one to find any reliable structural model representing the Walsh rule (Krygowski and Turowska-Tyrk 1990). Quite the opposite situation was found for the dependence of geometrical parameters ($\triangle$-value) on σ_p^+-values for p-X-Ph-Y systems in which Y stood for an electron-accepting substituent and X for both electron-accepting and electron-donating ones. In this case, variation in $\triangle$-values was some five times larger than previously and neither precision nor thermal motion correction for geometry affected significantly the final relationship which took the form of a linear dependence between $\triangle$ and σ_p (σ_p^+) (Krygowski 1987).

Analysis of Geometry of Systems with Intramolecular Interactions.

Now let us consider a few examples in which precise structural data (molecular geometry) permit a conclusion which may help one understand better the relationships between structure and reactivity.

The geometry of p-nitroaniline has long been the subject of structural investigations (Trueblood, Goldish and Donohue 1961; Colapietro, Domenicano, Marciante and Portalone 1982). In most cases, even in recent studies (Domenicano et al. 1982), the analysis of structural patterns was interpreted in terms of resonance theory (canonical structures) in which only 1, 2 and 6 canonical structures were taken into account (Fig. 6).

Let us look at such a system once again. Fig. 7 presents the geometry of N,N-dimethyl-p-nitroaniline (abbreviated as DPNA) (Maurin and Krygowski 1988). By calculating the contributions of canonical structures by use of the HOSE-model (Krygowski, Anulewicz and Kruszewski 1983) applied to this geometry, one can easily find that structure 6 is the least important of all nine canonical structures, presented in Fig. 6.

Table 2 presents the weights of contributions 1 - 9 of Fig. 7 for 12 different p-substituted derivatives of nitrobenzene (Krygowski and Turowska-Tyrk 1989).

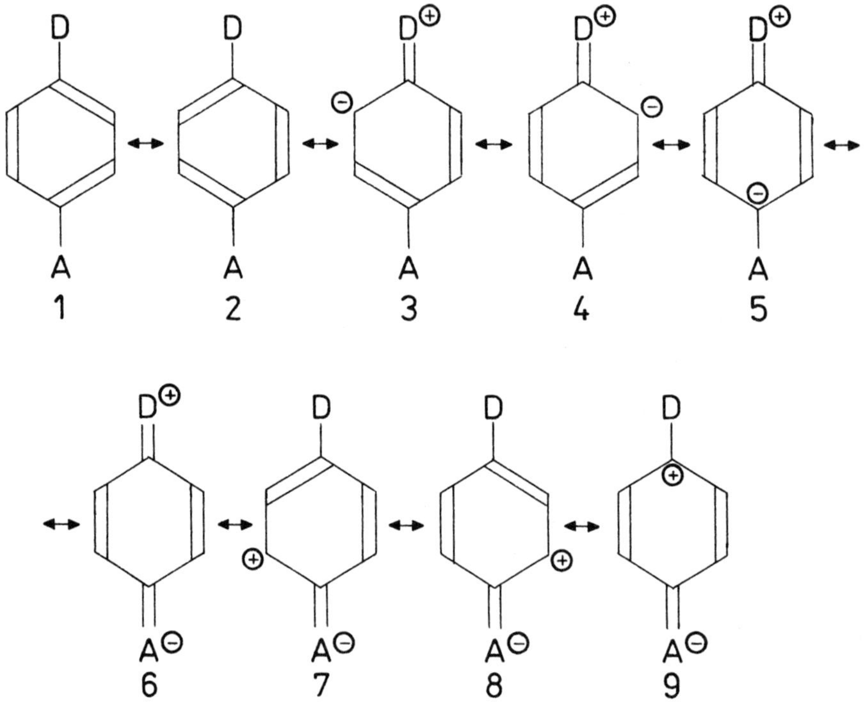

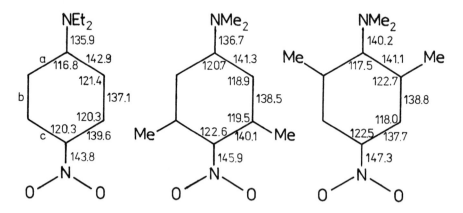

Fig. 6. Canonical structures for p-D-Ph-A type molecules
(A - electron accepting, D - electron donating
substituents).

Fig. 7. Geometry of: N,N-diethyl-4-nitroaniline (Maurin
and Krygowski 1988), N,N-dimethyl-4-nitro-3,5-
xylidine (Krygowski and Maurin 1989) and N,N-
dimethyl-4-nitro-2,6-xylidine (Maurin and
Krygowski 1987). Bond angles in degrees.

Table 2. Weights of canonical structure contributions (1 - 9), Fig. 7 for eleven p-XPhNO$_2$. For refs, cf. Krygowski and Turowska-Tyrk (1990).

pXPhNO$_2$ X =	Contribution of canonical structures 1 + 2	3 + 4 + 5	6	7 + 8 + 9	σ_p or σ_p^+
$\overline{O}$I$^-$	18.4	47.7	11.7	22.2	-2.9
NEt$_2$ A	20.7	47.3	10.3	21.7	-1.67
NEt$_2$ B	22.1	44.9	10.1	22.9	
NH$_2$	22.4	44.4	10.1	23.1	-1.47
-NHMe	24.0	42.6	9.7	23.7	-1.57
OH α	32.9	31.8	7.6	27.7	-0.91
OH β	33.1	31.8	7.6	27.5	
OCH$_3$	30.9	33.6	8.1	27.4	-0.79
COO$^-$ A	36.5	31.4	6.8	25.3	-0.41
COO$^-$ B	37.5	31.5	6.5	24.5	
Ph	35.4	33.4	6.9	24.3	-0.21
COOH	37.1	32.6	6.5	23.8	0.45
NO$_2$	40.8	26.6	6.0	26.6	0.78
NH$_3^+$	39.7	27.3	6.3	26.7	0.82

The substituted species are given in a sequence of decreasing electron donating power (increase in σ_p^+ or σ_p-values) and it is apparent that the weights of (1+2) increase in this direction whereas the weights of (3+4+5) and (6) decrease. This is in line with chemical expectation but more important seems to be the very low and not even approximately monotonic change of weights (7+8+9). This may be interpreted as follows: a decrease in electron-donating power of the substituent Y (an increase in σ_p^+ or σ_p) results in a decrease in the polarization of the ring by addition of electrons and hence causes a roughly monotonic decrease of weights (3+4+5) and (6) - and in line with this an increase of (1+2). However, the electron accepting demand of NO$_2$ does not react too strongly on this variation and π-electron polarization of the ring by NO$_2$ (i.e. presented by contributions (7+8+9)) is rather weak. This result is in line with an earlier view (Exner, 1966) that in systems like p-nitroaniline the major additional conjugation is that of the donating

substituent with the π-electron system of the ring
while the accepting substituent acts mainly by its
inductive effect. This point may be studied in more
detail by analysing the geometry of DPNA and its 2,6-
dimethyl and 3,5-dimethyl-derivatives. The geometry
of these systems is shown in Fig. 7 and the weights
of contributions (1÷9), Fig. 6, are collected in
Table 3.

Table 3. Weights of canonical structure contributions
(1 - 9), Fig. 6, for DPNA and its 2,6- and
3,5-dimethyl derivatives (in %).

Structure	Canonical structure contribution			
	1 + 2	3+4+5	6	7+8+9
NEt$_2$ structure A / B (Maurin and Krygowski, 1988)	20.7 / 22.1	47.3 / 44.9	10.3 / 10.1	21.7 / 22.9
2,6-dimethyl NEt$_2$ (Maurin and Krygowski, 1987)	32.2	37.0	7.3	23.5
3,5-dimethyl NMe$_2$ (Krygowski and Maurin, 1989)	24.4	44.5	8.9	22.2

From these data it is evident that there is a great
difference between canonical structure weights of the
2,6-dimethyl derivative of DPNA and DPNA itself. That
is due to the out-of-plane torsion of NEt_2 plane by
60.4° (Krygowski and Maurin, 1989) - steric hindrance
for π-conjugation between the $NAlk_2$-substituent and
the ring is responsible for a dramatic change in canonical structure distribution. If we look at DPNA
and its 3,5-dimethyl derivative, the NO_2-group is
twisted by 50.6° (Maurin and Krygowski, 1987) but, in
spite of the steric hindrance, the canonical structure
weights for DPNA and its 3,5-dimethyl derivative are
quite similar. The same conclusion may be drawn when
applying statistical analysis to the differences between individual bond lengths of those three systems
(Krygowski and Maurin, 1989). Evidently, these results are contrary to the classical view established
for descriptions of p-nitroaniline-like systems by
the dominating amount of contribution (6). Thorough
VB-calculations for p-nitroaniline and p-nitrophenol
are in line with these conclusions (Ohanessian and
Hiberty, 1984).

Another analysis of such systems is also possible
by simply applying the Walsh-Bent rule and the Δ vs
α plot. Examination of the geometry of DPNA suggests
that both central bonds (b) due to the mesomeric effect should become shorter whereas both pairs of a and
c bonds should become longer. We can see that the
length of a > c in a statistically significant way
($|a - c| > 3\sigma$). Why do we meet such a discrepancy? The
answer is quite simple: the geometry (bond lengths,
this time) is ruled not only by π-electron interactions (mesomeric effects) but also by σ-electron
effects. Since electronegativity of $NMe_2 \ll NO_2$ (χ values are 2.40 and 4.83, respectively (Huheey 1965,
1966)), we should expect the change in bond length
predicted by the Walsh rule. This applies to p-sym-
metrically disubstituted benzene derivatives, as shown
by the continuous line in Fig. 8 (Krygowski, 1984).
It is clear that the distance from the line $\delta\Delta(NEt_2) <
\delta\Delta(NO_2)$ and, since $\Delta = b - a$ and b depends only
slightly on the nature of substituent $b(NO_2) \approx b(NMe_2)$,
hence $a(NO_2) < a(NMe_2)$.

As a result we may conclude as summarized in
Fig. 8.

Another very useful application of molecular
geometry is in the analysis of steric effects. In most
cases steric hindrance is studied by comparison of
structural parameters like torsion or dihedral angles
(cf. e.g. Fujisawa, Oonishi, Aoki and Ohashi 1986) and
a very qualitative discussion of enlargement of bond

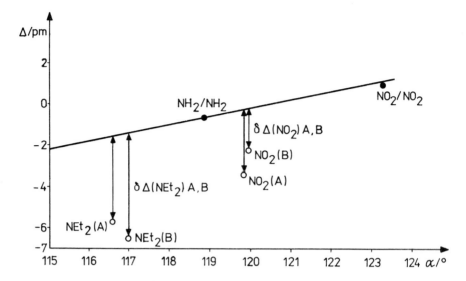

Fig. 8. Graph of bond length difference, Δ , vs angular change, α , for di-nitrobenzene (NO_2 NO_2),
phenylenediamine (NH_2 NH_2), full circles, and
N,N-diethyl-4-nitroaniline, open circles.

angle as a result of in-plane deformation. In order to
study interactions of this kind more quantitatively
let us define a few useful parameters of deformation
(Krygowski, Anulewicz, Daniluk and Drapała, 1990).
Consider for simplicity an example of an ortho
disubstituted benzene derivative with substituents R_1
and R_2, schematically presented in Fig. 9.

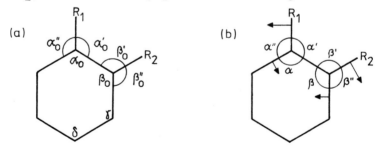

Fig. 9. Assignments of angles (degrees) in o-disub-
stituted derivative of benzene. Angles with
zero subscript refer to those for which steric
effects are not taken into accounts (a). Arrows in (b) represent forces of repulsion
whereas angles without subscript indicate the
real values of angles.

If there are no steric interactions between substituents R_1 and R_2 then $\alpha_0' = \alpha_0''$ and $\beta_0' = \beta_0''$. The values of α_0', β_0', α_0'' and β_0'', in general φ_0' or φ_0'', may easily be obtained from the Domenicano - Murray-Rust (1979) angular substituent parameter, $\triangle \varphi$.

By definition

$$\varphi = 120 + \triangle\varphi$$

and in our case we assign

$$\varphi_0 = 120 + \triangle\varphi$$

to indicate that φ_0 is estimated by assumption of the absence of any steric interactions. Only the additive scheme of electronic interaction between the substituent and the ring is taken into account. This procedure may fail to some extent if substituents R_1 and R_2 strongly interact electronically (by the through-resonance effect). If, however, these interactions are negligible (or if we neglect them although aware of their existence), we may calculate

$$\varphi_0' = \frac{360 - \varphi_0}{2}$$

and do the same for φ_0''.

We expect that φ', φ'' and φ may be deformed by repulsive forces exerted by both o-substituents R_1 and R_2.

Thus we define a Repulsive Deformation Parameter, RDP, of an angle φ' or φ'' (i.e. α', α'', β' and β'' in our case) by the formula

$$\text{RDP}\,(\varphi' \text{ or } \varphi'') = \varphi'(\text{or } \varphi'') - \varphi_0$$

As a result of deformations of φ' and φ'' it may be expected that these deformations, due to their relaxation, may change , by reason of non-additivity of angular substituent effects at both o-carbon atoms. These effects have been observed (Więckowski and Krygowski 1985, Grabowski and Krygowski 1985) and presented as the non-additivity parameter, NAP

$$\text{NAP} = \varphi - \varphi_0$$

For planar systems RDP(φ') + RDP(φ'') + NAP (φ) = 0

If the above parameters, RDP or NAP, exceed 3σ in absolute value, for a given angle, then we accept them as significant (at the level 0.0027).

Let us consider, as an illustration, the geometry of 2,5-dinitrobenzoic acid (Grabowski and Krygowski 1985), presented in Fig. 10.

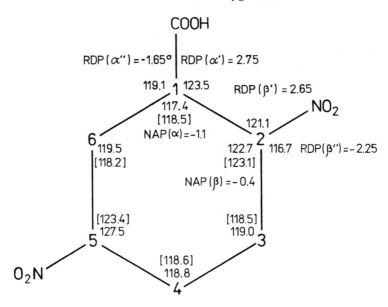

Fig. 10. Geometry of 2,5-dinitrobenzoic acid (Grabow-
          ski and Krygowski, 1985), showing also RDP-
          and NAP-values. Angles in degrees.

In this case we have R_1 = COOH and R_2 = NO_2 and the
application of angular substituent parameters leads
to α_0 = 118.5° and β_0 = 123.1°. Hence, for α_0'
and β_0', one finds 120.75° and 118.45°, respectively.
Now we may calculate RDP(α') = 123.5 - 120.75 = 2.75°
and RDP(β') = 2.65°; the RDP-values for α'' and β'' are
- 1.65° and - 2.25°, respectively. The negative RDP
means decrease as a result of deformation of φ'' -
angles, in line with the scheme of repulsive forces
acting between R_1 and R_2, presented as arrows in
Fig. 9.
        The RDP-value estimates numerically the magnitude
of deformation of an individual angle. If one wishes
to know what is the magnitude of deformation of all
angles at a given i^{th} atom, then another deformation
parameter must be defined: a Local Atomic Angular
Deformation Parameter, LAADP:

$$LAADP(i) = \left| RDP(\varphi_i') \right| + \left| RDP(\varphi_i'') \right| + \left| NAP(\varphi_i) \right|$$

By adding absolute values we obtain the quantity representing the sum of all deformations.
        Sometimes it is reasonable to analyze the de-
formations of some particular angles in the region of

overcrowding in the molecule. If we take into account only φ'-angles (in our case in Fig. 8 α' and β'), we may define a Global Angular Deformation Parameter

$$GADP(\varphi') = \sum_{i=1}^{2} RDP(\varphi'_i)$$

However, in some cases one wishes to estimate the global deformation of all angles of atoms in the region with overcrowding in the molecule. Then the Global Deformation Parameter of the Overcrowded Region is defined

$$GDPOR = \sum_{i=1}^{2} LAADP(\varphi_i)$$

The same may be obtained by summing up GDP-values over all the angles φ, φ' and φ''. All these parameters are convenient quantitative measures of structural deformations due to steric effects. To illustrate their utility, Table 4 presents some of them for several o,o'-disubstituted biphenyl derivatives, whereas Fig. 11 presents a scatter plot of GDPOR vs Charton's steric constant υ (Charton 1985).

Table 4. Structural deformation parameters describing steric effects on angular geometry the overcrowded region of a several o,o'-disubstituted derivatives of biphenyl.

$X_1=X_2$ (υ)	ψ^{**}	φ	RDP(i) i= 1	2	3*	4*	GDP
Cl 0.55	69.2°	φ'_i	1.45	1.15	–	–	5.2
		φ''_i	-0.95	-1.05	–	–	4.0
		φ_i	-0.5	-0.1	–	–	1.2
		LADP	2.9	2.3	–	–	10.4
NH_2 0.35	58.2°	φ'_i	0.1	1.15	–	–	2.5
		φ''_i	0.4	-1.75	–	–	4.3
		φ_i	-0.5	0.6	–	–	2.2
		LADP	1.0	3.5	–	–	9.0
COOH 1.39	71.3°	φ'_i	4.15	3.45	3.45	3.65	14.7
		φ''_i	-2.95	-4.35	-4.55	-1.55	13.4
		φ_i	-1.2	0.4	0.8	-2.2	4.6
		LADP	8.3	8.2	8.8	7.4	32.7

COOH	83.6°	ϱ'_i	5.45	3.55	3.75	4.55	17.3
1.39		ϱ''_i	-3.35	-4.65	-4.65	-2.05	14.7
		ϱ_i	-2.1	0.8	0.5	-2.5	5.9
		LADP	10.9	9.0	8.9	9.1	37.9
OCH_3	81.1°	ϱ'_i	-3.15	0.85	0.85	-3.15	8.0
0.36		ϱ''_i	3.55	-1.55	-1.55	3.55	10.2
		ϱ_i	-0.4	0.7	0.7	-0.4	2.2
		LADP	7.1	3.1	3.1	7.1	20.3
F	54.7°	ϱ'_i	3.1	-0.85	-0.55	2.7	7.2
0.27		ϱ''_i	-0.4	-1.85	-1.95	-0.2	4.4
		ϱ_i	-2.8	2.7	2.4	2.5	10.4
		LADP	6.3	5.4	4.9	5.4	22.0
Ph	42.1°	ϱ'_i	2.95	2.35	0.15	-0.9	6.35
0.57		ϱ''_i	-2.25	-3.25	1.15	1.3	7.95
		ϱ_i	-0.7	0.8	1.2	-0.4	3.1
		LADP	5.9	6.4	2.5	2.6	17.4
Ph	62.1°	ϱ'_i	2.35	2.95	-0.25	0.7	6.25
0.57		ϱ''_i	-3.25	-2.25	-0.05	-0.1	5.65
		ϱ_i	0.8	-0.7	0.1	-0.7	2.3
		LADP	6.4	5.9	0.4	1.5	14.2
COOH	54.4°	ϱ'_i	-3.0	1.65	3.85	4.95	13.45
1.39°		ϱ''_i	3.8	-2.55	-3.9	3.85	14.1
OCH_3		ϱ_i	-0.8	0.7	-0.2	1.1	2.8
0.36°		LADP	7.6	4.9	7.95	9.9	30.35

[*] Due to symmetry of the species RDP and LADP-values
are the same as for i=1 and 2. Hence the respective
values GDP are doubled sum over i=1,2.

[**] ψ stands for dihedral angle between the planes of
both phenyl rings in question.

For details cf. Krygowski, Anulewicz, Daniluk and Drapała (1990).

It seems reasonable that υ -values, which are often
used to study the dependence of chemical reactivity
on spatial requirements of sterically interacting substituents, may also be used to analyze the deformations
of structure due to steric interactions.
     It should be noted that the RDP-procedure and all
the deformation parameters presented may be applied
only to benzene derivatives since the angular substituent parameters were estimated only for these systems.

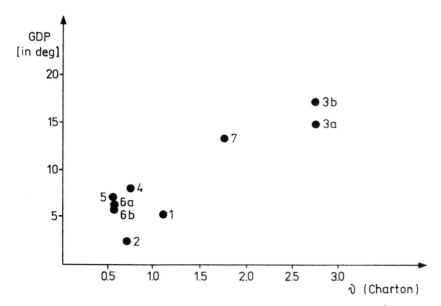

Fig. 11. Dependence of GDP-values on Charton's steric
parameter for 9 o,o-disubstituted diphenyl
derivatives (Krygowski, Anulewicz, Daniluk
and Drapała, 1990).

However, in other systems, built of six-membered aromatic rings (including condensed benzenoid hydrocarbons as well as heteroaromatic systems), one may replace 120° by the φ_o-value taken from the geometry
of the unsubstituted system. In this way, we implisit-
ly accept an assumption that deformation due to substitution is of the same kind and of the same numeric-
al value, $\Delta\varphi$, as for the benzene ring. Another, even
simpler procedure for such cases could involve taking
φ instead of φ_o, viz.

$$RDP(\varphi' \text{ or } \varphi'') = \varphi'(\text{or } \varphi'') - \frac{360° - \varphi}{2}$$

We should, however, be aware that, in this case, no
internal substituent effect is taken into account.

Conclusion.

Precise geometry of chemical species determined
by X-ray diffraction may be applied successfully to
analyze inter- and intra-molecular interactions,
provided the analysis is carried out with care and
by using properly defined models.

Acknowledgement

This work was supported by the Project RP.II.10 from the Ministry of National Education.

References.

Bernstein J. (1985) Lecture Notes of 11th Course of International School of Crystallography, Erice 24 May – 6 June, 1985, CNR Reparto Tecnografico, Roma 1985, p. 268 ff.

Bondi A. (1964) J. Phys. Chem., 68, 441–455.

Bürgi H.-B., Dunitz J.D. (1983) Acc. Chem. Res., 16, 153–164.

Dashevsky V.G., Konform$_a$tsyonnyi Analiz Organicheskikh Molekul (in Russian) Moscow, Izd. Khimiya, 1982, p. 94.

Domenicano A. and Murray-Rust P., (1979) Tetr. Letters, 2283–2286.

Domenicano A., Vaciago A. and Coulson C.A. (1975) Acta Cryst., B31, 221–234.

Exner O. (1966) Collect. Czechoslov. Chem. Comm., 31, 65–81.

Fujisawa S., Oonishi I., Aoki J. and Ohashi Y. (1986), Acta Cryst., C42, 1390–1392.

Grabowski S.J. and Krygowski T.M. (1985) Acta Cryst., C41, 1224–1226.

Hiberty P.C. and Ohanessian G. (1984) J. Amer. Chem. Soc., 106, 6963–6968.

Hoffmann R. (1983) foreward to the monograph by L.V. Vilkov, V.S. Mastryukov and N.I. Sadova: Determination of the Geometrical Structure of Free Molecules, Mir Publishers, Moscow 1983, p. 5.

Huheey J.E. (1965) J. Phys. Chem., 69, 3284–3292.

Huheey J.E. (1966) J. Phys. Chem., 70, 2085–2091.

Jaskólski M. (1986) Pol. J. Chem., 60, 263–271.

Kitaygorodzky A.I. (1973) Molecular Crystals and Molecules, Academic Press, New York

Krygowski T.M. (1984) J. Chem. Res., 238–239.

Krygowski T.M. (1987) J. Chem. Res., 120–121

Krygowski T.M., Anulewicz R., Daniluk T. and Drapała T., Struct. Chem., in press.

Krygowski T.M., Anulewicz R. and Kruszewski J. (1983) Acta Cryst., B39, 732-741.

Krygowski T.M. and Maurin J. (1989) J. Chem. Soc. Perkin II, 695-698.

Krygowski T.M. and Turowska-Tyrk I., Pol. J. Chem. (1990) in press.

Krygowski T.M. and Turowska-Tyrk I. (1990) Coll. Czechoslov. Chem. Comm. in the press and poster at ISNA-6, Osaka 20-25.VIII.1989.

Maurin J. and Krygowski T.M. (1987) J. Mol. Struct., 158, 359-366.

Maurin J. and Krygowski T.M. (1988) J. Mol. Struct., 172, 413-421.

Murray-Rust P. (1987) in Molecular Structure and Biological Activity (Editors Griffin J.F. and Duax W.L.) Elsevier pp. 117-133.

Pertsin A. and Kitaygorodzky A.I. (1987), The Atom-Atom Potential Method, Springer Verlag, Berlin.

Trueblood K.N., Goldish E. and Donohue J. (1961) Acta Cryst., 14, 1009-1017.

Turowska-Tyrk I., Krygowski T.M., Gdaniec M., Hafelinger G. and Ritter G. (1988) J. Mol. Struct., 172, 401-412.

6

The effect of solvent–surface interactions on crystal growth and dissolution

L. J. W. Shimon, M. Vaida, L. Addadi, M. Lahav, and L. Leiserowitz

ABSTRACT

The effect of solvent on crystal growth and dissolution studied by two stereochemical strategies is described[1]. The first approach involved the growth of N-(E-cinnamoyl)-(S)-alanine from acetic acid and of three crystalline hydrates (asparagine·H_2O, rhamnose·H_2O and S-lysine·HCl·$2H_2O$) in the presence of primary alcohols as "tailor-made" solvents. It was demonstrated that stereospecific adsorption of the tailor-made solvent at a particular face inhibits its growth. The second approach, concerned the growth and dissolution of hemihedral faces at the opposite ends of the polar axes of (R,S)-alanine and γ-glycine in water, methanol and ethanol, one pole exposing carboxylate groups and the other amino groups. Relatively fast growth and dissolution at the carboxylate end in water was explained in terms of strong solvent binding at a subset of surface sites and solvent repulsion at the remaining sites, so allowing for easy access of solute at the latter unhydrated surface sites followed by expulsion of water at the former. The repetitive succession of these steps results in a "relay" type mechanism of crystal growth.

1.   INTRODUCTION

Solvent has a strong influence on the structure and habit of crystalline materials; however, the role played by the solvent-surface interactions in enhancing or inhibiting crystal growth is still not well understood. To date there have been two distinctly different approaches to clarify this point. Calculations based on "surface-roughening" considerations predict that favorable interactions between solute and solvent on specific faces will lead to reduced interfacial tension, causing a transition from a smooth to a rough interface, and a concomitant faster surface growth.[2,3] Alternatively, it has been proposed that the preferential adsorption of solvent molecules at specific faces will inhibit growth of those faces as removal of a bound solvent molecule poses an additional energy barrier for continued growth.[4,5,6,7] In some of these studies, the binding of polar solvents was estimated by calculations of the electrostatic potential at the crystal surface and interpreted in terms of crystal surface hydrophobicity and philicity.[5,6]

We have previously reported that stereospecific adsorption of "tailor-made" additives effects both the growth and dissolution of the crystal surfaces to which the additive can bind, the absorbate being a substrate molecule with an altered moiety[8,9]. This adsorption was manifested by changes in crystal morphology with the rule that binding of the additive to a specific face causes an inhibition of growth perpendicular to that face generally resulting in an increase of its surface area relative to that of unaffected faces. Here we shall probe the assumption that tight binding of solvent will be analogous to the binding of additives. Two possible strategies to pin-point solvent-surface interactions and thereby clarify the solvent effect have been adopted. The first involves the growth of crystalline non solvates and solvates in the presence of tailored solvents which have been targeted for stereospecific interactions at particular crystal surfaces. The selective adsorption of these modified solvents during crystal growth provides a means for isolating solvent-surface interactions. The second method involves the use of materials with polar axes for a study of relative rates of growth of hemihedral crystal faces in a variety of solvents. In such polar materials we exploit the fact that differences in the growth rates at the opposite ends of the polar axis arise primarily from differences in solvent-surface interactions.

2. RESULTS AND DISCUSSION

2.1 Crystal growth in the presence of tailor-made solvents

1a   **1b**   **1c**

N-(E-cinnamoyl)-(S)-alanine The molecules of N-(E-cinnamoyl)-(S)-alanine, *1a*, crystallize in spacegroup $P2_1$, (a=6.1, b=8.2, c=11.3Å, β=90.6 °). The packing arrangement[10] delineated by the crystal faces as grown from methanol solution is shown in Fig. 1. The molecules are arranged such that the carboxyl groups emerge at the $\{1\bar{1}1\}$ faces and the C(chiral)-H bonds are directed along the $+b$ axis. In a study on the effect of tailor-made additives on the morphology of pure N-(E-cinnamoyl)-(S)-alanine (Fig. 2a), it was found[10] as expected that the methyl ester *1b*, of the same absolute configuration as *1a*, induced large $\{\bar{1}\bar{1}\bar{1}\}$ faces (Fig. 2b). Tailor-made additive N-(E-cinnamoyl)-(R)-alanine *1c*, of configuration opposite to that of the host induces, as expected, formation of an (010) face (Fig. 2c)[10]. Acetic acid is a solvent which can selectively bind at the exposed carboxylate groups of the $\{\bar{1}\bar{1}\bar{1}\}$ faces forming a hydrogen bonded dimer *2a*. Acetic acid can also bind to the **CHCO$_2$H moiety of *1a* via a cyclic dimer *2c* on the (010) face. The dimer *2b* is the motif adopted by acetic acid in its own crystal structure[11], forming an O-H···O bond and a C-H···O(carbonyl) interaction[12]. Crystallization of *1a* from glacial acetic acid yields crystals with the morphology shown in Fig. 2d, in keeping with expectation.

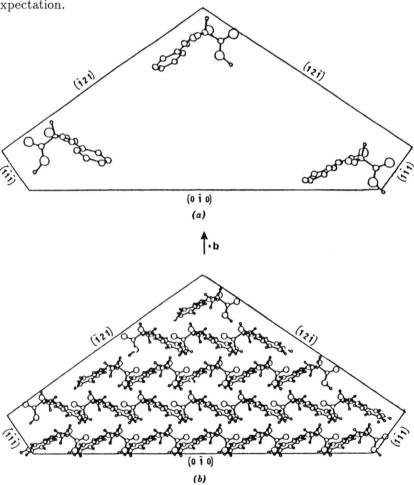

Fig. 1   Packing arrangement of N-(E-cinnamoyl)-(S)-alanine delineated by the faces observed in the pure crystal: (a) for clarity, only three molecules shown; (b) complete arrangement.

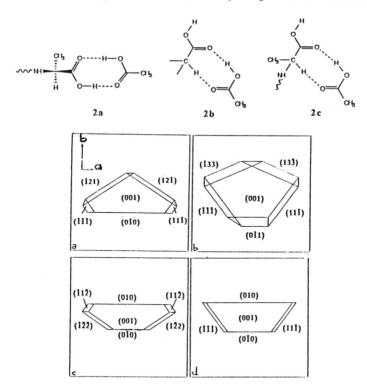

Fig. 2 Computer drawn pictures of measured crystals of N-(E-cinnamoyl)-(S)-alanine; (a), (b), (c) refer to crystals grown from methanol, (a) pure crystal; (b) grown in the presence of the methyl ester *1b*, (c) grown in the presence of E-cinnamoyl-(R)-alanine *1c*; (d) pure crystal grown from acetic acid.

(S)-Asparagine monohydrate To facilitate the transition from additive-surface to solvent-surface interactions, crystalline hydrates were chosen so that added solvents, such as methanol and higher alcohols, may be considered as both "tailor-made additive" and "tailor-made solvent". A systematic solvent-dependent change in morphology has been carried out using as substrate the resolved conglomerate (R) or (S) asparagine monohydrate. $(NH_2COCH_2CH(NH_2)COOH \cdot H_2O$, $a=5.584$, $b=9.735$, $c=11.701\text{Å}$, space group $P2_12_12_1$, $Z=4)$ [13]. This crystal grown in aqueous solution exhibits as many as 18 faces (Fig. 3a) of the families {011}, {010}, {101}, {012}, and {111}[14]. We expected linear alcohols (e.g. methanol) to behave as a tailor-made additive and be adsorbed primarily on the {010} or the {011} face, replacing H_2O with a hydroxyl hydrogen bond and oriented so that its alkyl group emerges from these faces (Fig. 3c). When the water in the aqueous solution is gradually replaced by either methanol or ethanol, there is a

marked increase in the relative surface area of the {010} crystal faces with respect to the original morphology (Fig. 3b), the crystal eventually becoming platelike. This increase in size of the {010} face is indicative of an inhibition of growth perpendicular to it and is akin to the changes in morphology of asparagine which have been affected by "tailor-made" additives[15,16] The arrangement of the hydrate molecules is such that one of the two O-H bonds of each water molecule emerges from the {010} face and the other points into the crystal bulk, which means that in principle all four symmetry related hydrate water molecules can be replaced by alcohol. On the {011} face only two out of the four symmetry-related water molecules can be substituted by methanol and inhibition is less dramatic. That methanol can indeed be selectively absorbed on both the {010} and {011} faces was further demonstrated by experiments involving initial dissolution of asparagine·H_2O crystals in methanol solution, revealing etchpits on only these faces, although poorly developed.

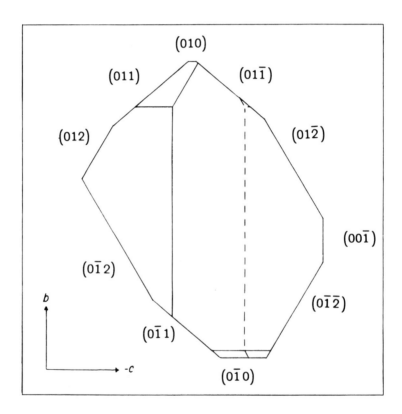

Fig. 3(a)   Morphology of a crystal of (S)-asparagine monohydrate grown from aqueous solution, as viewed down the *a* axis.

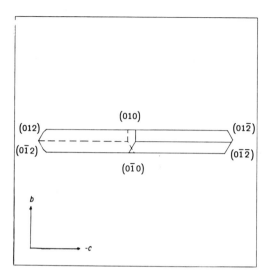

Fig. 3(b)   Morphology of a crystal of (S)-asparagine monohydrate grown from 30:70 methanol:water solution, as viewed down the *a* axis. Note, the inhibition of growth along the *b* axis.

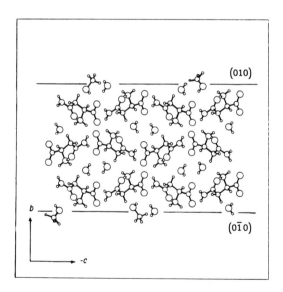

Fig. 3(c)   Packing arrangement of (S)-asparagine monohydrate viewed down the *a* axis. The hydrate waters are oriented with one of their two OH bonds emerging from the {010} faces. Replacement of these water molecules by methanol is depicted.

Rhamnose monohydrate In rhamnose monohydrate $(C_6H_{12}O_5, a=7.901, b=7.922, c=6.670 \text{Å}, \beta=95.52°$, space group $P2_1, Z=2)^{17}$ advantage was taken of the existence of the polar b axis which allows for a direct comparison of affected and unaffected faces on the same crystal. The hydrating water molecules are oriented with their O-H bonds pointing in the $+b$, but not $-b$, direction (Fig. 4a). Thus in the presence of alcohols we expected changes from the regular crystal morphology at the $+b$ side of the crystal, but not at the $-b$ side. Upon addition of CH_3OH in the solution, we observed a decrease in the relative rate of crystal growth in the $+b$ as against $-b$ direction *vis a vis* that of the crystal grown in aqueous solution, indicating a relative inhibition of growth at the $+b$ end. Furthermore, there is a pronounced increase in the size of $\{100\}$ faces as well as that of the $\{110\}$ faces relative to that of $\{1\bar{1}0\}^{18}$ faces (Fig. 4b and c). This inhibition can be understood in terms of the relative ease of binding alcohol to these faces. According to figure 4a, methanol can be easily adsorbed at the two different water sites on the $\{110\}$ face, each hydroxyl group participating in three hydrogen bonds and the side-chain being forced to emerge from the $\{110\}$ face. On the opposite hemihedral $\{1\bar{1}0\}$ faces, only one of the two different water sites can be replaced by alcohol with the hydroxyl group bound by only one hydrogen bond (Fig. 4d). The pronounced increase in the area of the $\{100\}$ faces is compatible with the observation that on this face both water sites can be replaced by alcohol with the side chain emerging from the surface.

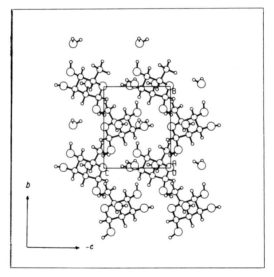

Fig. 4(a)   Packing arrangement of α-rhamnose monohydrate viewed down the a axis. The OH bonds of the hydrate waters point towards the $+b$, but not the $-b$, direction

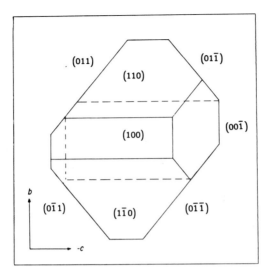

Fig. 4(b) Morphology of a crystal of α-rhamnose monohydrate grown from aqueous solution, viewed down the *a* axis.

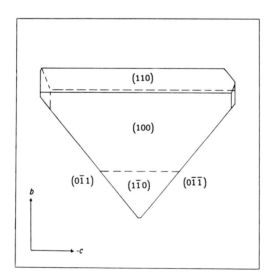

Fig. 4(c) Morphology of α-rhamnose grown from 90:10 methanol:water solution, viewed down the *a* axis. Note the marked change in the morphology at the +*b*, but not -*b*, end of the crystal.

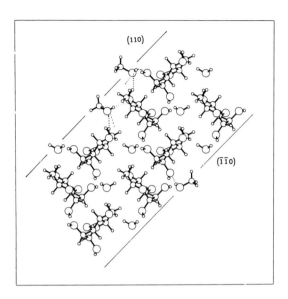

Fig. 4(d)   Packing arrangement of α-rhamnose monohydrate viewed down the *a* axis. The two different hydrate waters oriented with their OH bonds emerging from the (110) face are shown replaced by methanol. On this surface both water molecules may be replaced by methanol, which are each tightly bound via three hydrogen bonds (because of the projection only two hydrogen bonds are visible). On the ($\bar{1}\bar{1}0$) face methanol can replace only one of the two water molecules and which may be loosely bound by only one hydrogen bond.

Lysine hydrochloride dihydrate (S)-Lysine·HCl crystallizes from water as a dihydrate in a monoclinic structure of spacegroup $P2_1$ $(CO_2^-(NH_3^+)CH(CH_2)_4NH_3^+$ Cl⁻·2H_2O, $a=7.49$, $b=13.32$, $c=5.88$Å, $\beta=97.50$ °, $Z=2)$[19]. The crystals are prismatic in form with a triangular cross-section. The packing arrangement delineated by the crystal faces forming the triangle is shown in Fig. 5a. The lysine side chains are aligned parallel to the *b* axis with the ⁺H_3NCHCO_2⁻ moiety emerging from the well-developed {110} faces[10], the hemihedral {1$\bar{1}$0} faces are small in comparison. When grown in a 3:1 methanol-water solution the crystal assumes a more symmetric, truncated rhombic morphology, with increased development of the {1$\bar{1}$0} faces (Fig. 5b). This result appears to be consistent with inhibition of growth at the {1$\bar{1}$0} faces, as imposed by absorbed methanol thereon. We are left with the question, however, why methanol was not adsorbed in the pockets at the (0$\bar{1}$0) face, which would have left the triangular cross-sectional crystal shape unchanged. The pronounced development of the (0$\bar{1}$0) basal face from aqueous solution seems, at first sight, unusual because the face is composed of what may be regarded as an array of parallel molecular dipoles at the same level and so less stable than the less developed {1$\bar{1}$0} faces. We may explain this large (0$\bar{1}$0) face as resulting from strong solvent water–surface interactions. If so we may present an alternative or perhaps additional mechanism for the symmetric morphology (Fig. 5b) in so far that methanol does not stabilize the (0$\bar{1}$0) face, so leading to the development of the {1$\bar{1}$0} faces.

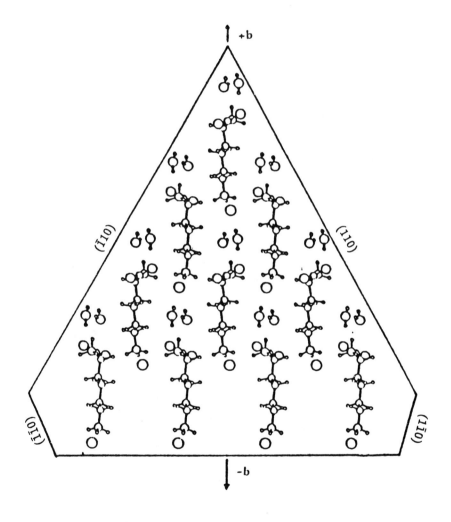

Fig. 5(a) Packing arrangement of the crystals of (S)-Lysine·HCl·2H$_2$O viewed along the *c* axis as delineated by the observed {*hk0*} crystal faces

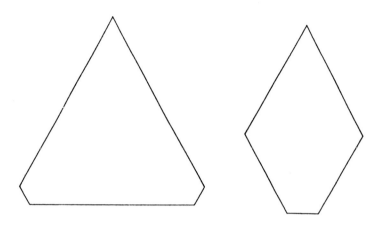

Fig. 5(b)   Computer-drawn pictures of typical crystals of (S)-Lysine·HCl·2H₂O viewed along the *c* axis grown in; (a) aqueous solution; (b) 3:1 CH₃OH-H₂O.

2.2 A "relay" mechanism for the crystal growth in (R,S)-alanine and γ-glycine.

The two hydrate systems aparagine·H₂O and rhamnose·H₂O discussed above demonstrate that the extension from tailor-made additives to tailored solvents is valid. We now examine polar axial systems without solvent of crystallization where it is generally difficult to determine at which surface site preferential binding of solvent occurs. A benchmark study on the effect of solvent on growth of polar crystals was carried out by Wells in 1949[4]. He found that in aqueous solution the α-form of resorcinol, space group $Pna2_1$, grows unidirectionally along the polar *c* axis. The crystal . (Fig. 6) exhibits what may be regarded as proton donor acidic {011} faces at one end of the crystal and proton acceptor basic {0$\bar{1}$1} faces[5] at the opposite end. The absolute direction of growth of resorcinol in aqueous solution was shown to take place at the hydroxyl-rich proton acceptor {0$\bar{1}$1} faces. The explanation for this effect appears to be inconclusive[5,7] perhaps because the calculated interactions between water and the two opposite faces were not sufficiently different.

Thus we have selected the two polar crystal systems of γ-glycine (space-group $P3_2$, or $P3_1$, $a=b=6.975$, $c=5.473$Å, $Z=3$)[20] and (R,S)-alanine (space-group $Pna2_1$, $a=12.06$, $b=6.05$, $c=5.82$Å, $Z=4$)[21] for their molecular simplicity, ability to form strong hydrogen bonds with water as well as the fact that they expose NH₃⁺ proton donors at one end of the polar *c* axis of the crystal and CO₂⁻ proton acceptors at the opposite end. These two structures have remarkably similar packing and morphological features (Figs. 7a, b, c

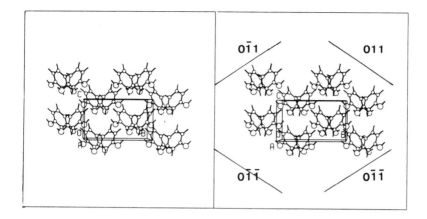

Fig. 6 Stereoscopic view of the packing arrangement of α-resorcinol.

and d), with an $(00\bar{1})$ face perpendicular to the polar c axis at one end of the crystal and capped faces at the opposite end. According to crystal growth and etching experiments on these two crystal systems [22a,b] as well as a Bijvoet x-ray structure analysis on γ-glycine[22b], the CO_2^- groups are exposed at the $(00\bar{1})$ face, the "flat -c end", while the NH_3^+ amino groups are exposed at the $+c$ capped end[22c]. The crystal growth and dissolution experiments also indicated that in aqueous solution the -c carboxylate end of crystals of γ-glycine and (R,S)-alanine grow and dissolve faster than the $+c$ amino end (Figs. 8a, b and c). The question remains as to which end of the crystal water may bind more tightly, and to correlate between the macroscopic phenomena with the binding and recognition of the solvent at a molecular level.

Inspection of the packing arrangement of both crystal structures (Fig. 7c,d) reveals that the $(00\bar{1})$ carboxylate faces comprise regular pockets on a molecular level and can be regarded as corrugated in two dimensions. This surface arrangement is generated by the screw axis perpendicular to the $(00\bar{1})$ face, which relates nearest-neighbor amino acid molecules and is enhanced by the fact that the molecules are aligned with their long axis parallel to the polar screw axis. (R,S)-alanine, because of the 2_1-axis, has only two levels of molecules on the $(00\bar{1})$ face, each laying $c/2=2.9$Å above or below its neighbor so fixing the depth of the pocket; in γ-glycine the pocket is deeper because of the 3_2 axis leading to a difference in level of $2c/3=3.6$Å. The faces which cut the amino end of (R,S)-alanine or γ-glycine, $\{201\}$, $\{011\}$ and $\{10\bar{1}3\}$, respectively, are comparatively smooth.

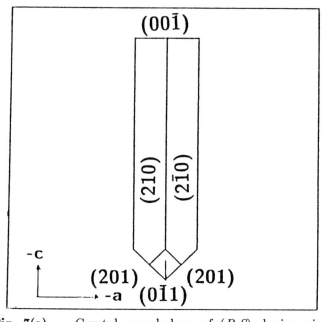

Fig. 7(a) Crystal morphology of (*R,S*)-alanine viewed down the *b* axis. The crystal exhibits nine faces of the type {210}, {011}, {201} and (00$\bar{1}$).

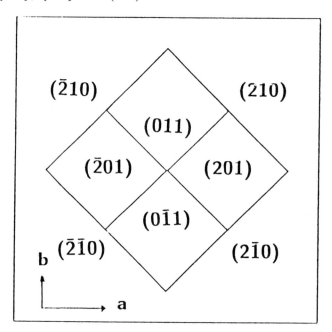

Fig. 7(b) Crystal morphology of (*R,S*)-alanine viewed down the -*c* axis.

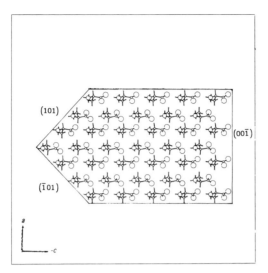

Fig. 7(c)   Packing arrangement of (R,S)-alanine delineated by crystal faces viewed down the b axis. The capped faces, $\{\bar{2}01\}$ and $\{011\}$ at the $+c$ end of the polar axis, expose NH_3^+ and CH_3 groups at their surfaces, the opposite $(00\bar{1})$ face exposes carboxylate CO_2^- groups. The two horizontal lines parallel to c do not represent faces seen edge on, but rather the intersections of the (210) and $(2\bar{1}0)$ faces and the $(\bar{2}\bar{1}0)$ and $(\bar{2}10)$ faces.

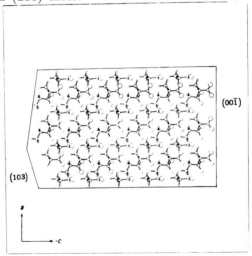

Fig. 7(d)   Packing arrangement of γ-glycine delineated by crystal faces viewed down the b axis. The capped faces $\{10\bar{1}3\}$ which cut the $+c$ axis exhibit different surface structures from that of the $(00\bar{1})$ carboxylate face. At the $+c$ end is shown the $(10\bar{1}3)$ face and the edge $2a + b + 1/3c$ which delineates the intersection between the $(\bar{1}103)$ and $(0\bar{1}13)$ faces.

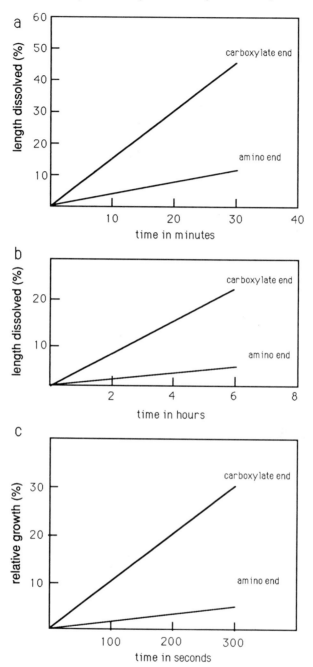

Fig. 8 (a) Graph of the dissolution of (*R*,*S*)-alanine in water.
(b) Graph of the dissolution of γ-glycine in water.
(c) Graph of the relative growth at the opposite poles of the polar axis of (*R*,*S*)-alanine in water.

We may explain the binding of water qualitatively as follows. The pockets at the $(00\bar{1})$ face expose primarily oxygen atoms, acting as proton acceptors for the NH_3^+ proton donor groups of the solute molecules and so fit to and bind the NH_3^+ moiety of the aminoacid (Figs. 9a, b, c and d). It has been previously established in numerous primary amide/carboxylic acid systems that when the N-H group of an N-H···O(carbonyl) hydrogen bond is replaced by a hydroxyl oxygen, the resulting (hydroxyl)O···O(carbonyl) lone-pair interaction will be repulsive[23]. More specifically, it was found that the molecules prefer to position themselves so that the O···O distance will be minimally 3.5Å, instead of 2.7 to 2.8Å for the N-H···O(carbonyl) distance.

Replacement of the NH_3^+ by water within the pockets of (R,S)-alanine and γ-glycine is possible in essentially two different orientations, one orientation comprises one O-H···O hydrogen bond and two O···O lone pair repulsions and the other two O-H···O bonds and one O···O lone pair-lone pair repulsion. Consequently introduction of water yields repulsive or at best weakly attractive interactions. The pocket will therefore be unhydrated or only slightly hydrated and relatively easily accessible to approaching solute molecules. In contrast, the water molecule may be strongly bound to the outermost layer of CO_2^- groups via O-H···O (carboxylate) hydrogen bonds[28]. As glycine or alanine molecules are incorporated into adjoining pockets, the CO_2^- groups of the newly added substrate molecules are within 3Å of the water bound on the outermost surface and expel the water thereby generating a new unsolvated pocket on the crystal surface. This relay process of solvent water binding and expulsion helps growth and dissolution by both desolvating the surface and perpetuating the natural corrugation of the surface, on a molecular level (see Figs. 10a and b). Conversely, the {011} and {201} faces of (R,S)-alanine which expose NH_3^+ and CH_3 groups at the $+c$ end of the crystal are relatively smooth and comprise molecules which are equally accessible for water binding. The corresponding three $\{10\bar{1}3\}$ faces of γ-glycine have a similar surface structure exposing NH_3^+ and CH_2 groups and their molecular surfaces should be equally accessible to water.

To further elucidate this proposed binding of water, we next chose a solvent which should be able to bind tightly to all sites on the NH_3^+ face almost as effectively as water but will be able to bind to all sites on the pocketed CO_2^- face as well; this solvent was methanol. The binding of methanol to the NH_3^+ groups at the capped crystal end may take place via its oxygen atom, but unlike water, methanol can bind within the pocket via a strong O-H···O (carboxylate) bond and three, albeit weak CH··O (carboxylate)[12] interactions, with length 3.1-3.3Å, as depicted in figures 9a and c.

Upon crystal growth in 20% methanol in water, (R,S)-alanine still grew faster at the carboxylate end than at the amino end, but at a smaller speed of growth relative to pure water. Indeed with 80%

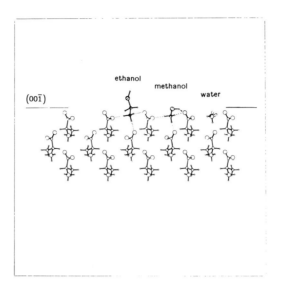

Fig. 9(a)   Packing arrangement of (R,S)-alanine viewed along *b* showing part of the exposed structure at the (001) face and molecules of H_2O, CH_3OH and CH_3CH_2OH adsorbed in the pockets. Note the lone pair-lone pair repulsion, between the water oxygen and the neighboring oxygens on the periphery of the pocket. Methanol makes CH⋯O contacts and an OH⋯O hydrogen bond in the pocket. Ethanol, although it makes CH⋯O contacts, cannot form an OH⋯O bond with the surface.

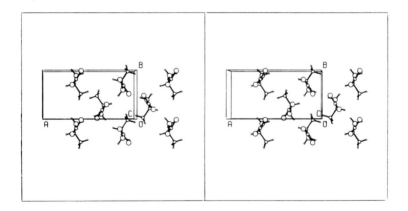

Fig. 9(b)   Stereoscopic view of the pocket on to the $(00\bar{1})$ face of (R,S)-alanine.

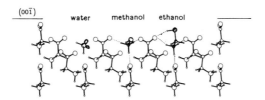

Fig. 9(c) Packing of γ-glycine viewed along b showing part of the exposed structure at the $(00\bar{1})$ face and molecules of H_2O, CH_3OH and CH_3CH_2OH adsorbed in the pockets. Note the lone pair-lone pair repulsion between the water oxygen and the neighboring oxygens at the periphery of the pocket. Methanol makes $CH\cdots O$ contacts and an $OH\cdots O$ hydrogen bond in the pocket. Ethanol is bound within the pocket by both $CH\cdots O$ contacts and also an $OH\cdots O$ hydrogen bond, which presence is possible because the pocket in γ-glycine is about 0.6Å deeper, than in (R,S)-alanine

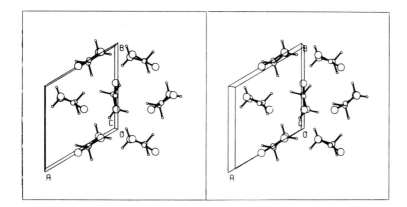

Fig. 9(d) Stereoscopic view of the pocket on to the $(00\bar{1})$ face of γ-glycine.

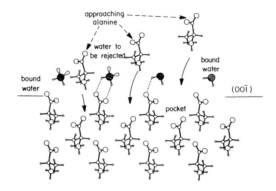

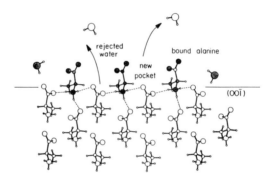

Fig. 10   Schematic representation of the (00$\bar{1}$) face of (*R,S*)-alanine during the crystal growth process.

(a) In this first view approaching solute alanine molecules are depicted about to be bound within the pockets of this face. Also shown are water molecules bound to the outermost layer of this face. The pockets remain primarily unsolvated because lone pair-lone pair oxygen-oxygen repulsion inhibits the binding of water within them.

(b) In this second view, at the (00$\bar{1}$) surface alanine molecules are bound via three NH⋯O hydrogen bonds. The previously bound water molecules are shown being rejected by O(water)⋯O(carboxylate) lone pair-lone pair repulsions. Note the formation of new unsolvated pockets.

methanol in water mixtures, the needle crystals grew slightly faster at the $+c$ amino end than at the $-c$ carboxylate end (Fig. 11a and b). Analogously, when crystals of γ-glycine or (R,S)-alanine were dissolved in methanol, their dissolution took place primarily from the $+c$ amino end of the crystal, the dissolution from the $-c$ carboxylate end having been inhibited by the binding of the methanol (Figs. 12a and 13a). These results are a clear indication that methanol is more strongly adsorbed at the $(00\bar{1})$ face than water. (R,S)-alanine grows as very thin [001] whiskers from sublimation; thus it is evident that both water and methanol inhibit growth along the c axis, in view of the dramatic relative increase in the ratio of the crystal area to its length[29]. The observation that the needlelike crystals of (R,S)-alanine grown in methanol/water mixtures are decidedly thinner than those grown in water suggests that methanol binds stronger than water to the $\{210\}$ side faces. These faces appear to have a more hydrophobic character than the $(00\bar{1})$ carboxylate and $\{201\}$ and $\{011\}$ amino faces as they expose $C-CH_3$ methyl groups[30] and it appears as if such molecules can be replaced by methanol where its OH group binds to a neighboring alanine molecule. The dissolution of these two remarkably similar systems proceeds differently in ethanol, giving further evidence that the structure of the surface plays a key role in the binding ability of the solvent. When (R,S)-alanine was dissolved in ethanol the crystals dissolved more rapidly from the carboxylate end (Fig. 12b). Growth of (R,S)-alanine from 78% ethanol in water showed analogously faster growth at the CO_2^- end relative to the NH_3^+ end (Fig. 11c). These results are understandable since the $(00\bar{1})$ face of (R,S)-alanine cannot bind ethanol as it is unable to form both $C-H\cdots O$ interactions and an $O-H\cdots O$ hydrogen bond within the pockets of (R,S)-alanine, leaving the $(00\bar{1})$ face free for growth or dissolution as in aqueous solution (Fig. 9a). On the other hand the γ-glycine pocket is sufficiently deep so as to allow ethanol binding via three methyl $CH\cdots O$ interactions as well as an $O-H\cdots O$ hydrogen bond (Fig. 9b). As expected, therefore, the dissolution of γ-glycine is inhibited from this direction (Fig. 13b). Finally if CF_3CH_2OH is used as a solvent, it can no longer bind within the CO_2^- pockets via $C-F\cdots O$ contacts, the solvent may in fact be even repelled from the CO_2^- crystal surface due to the net electronegative charge on the fluorine atoms. Indeed no inhibition of the dissolution of γ-glycine from the $-c$ end of the crystal is observed (Fig. 13c).

3. CONCLUSION

We have studied the effect of solvent on crystal growth on a molecular level and explained the phenomena in terms of molecular recognition at crystal surfaces.
Crystal growth experiments of N-(E-cinnamoyl)-(S)-alanine, asparagine monohydrate, rhamnose monohydrate and perhaps lysine HCl dihydrate demonstrate that an extension from "tailor-made" additives to solvents can be made. Namely, the "tailored" solvent will inhibit

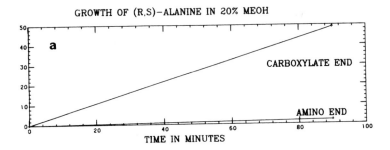

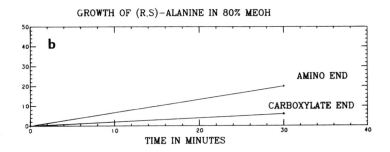

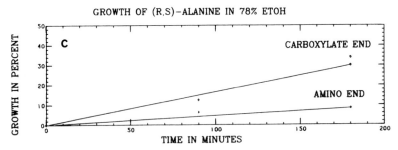

Fig. 11   Growth of (R,S)-alanine in (a) 20% methanol in water, in (b) 80% methanol in water and in (c) 78% ethanol in water.

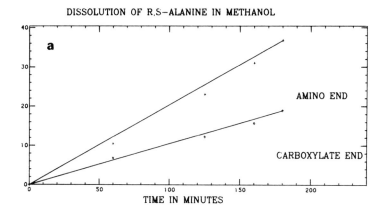

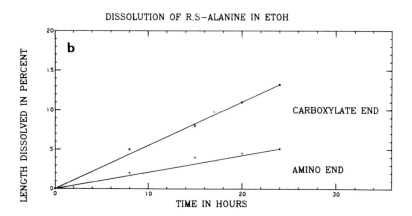

Fig.12 Dissolution of (R,S)-alanine in presence of (a) methanol and (b) ethanol.

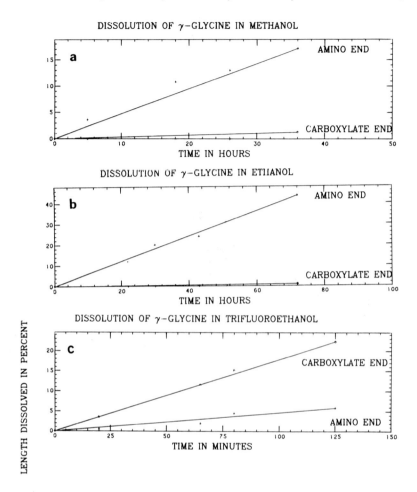

Fig.13   Dissolution of γ-glycine in presence of (a) methanol, (b) ethanol (Note the reversal of the direction of fastest dissolution compared to (R,S)-alanine) and (c) trifluoroethanol.

the growth of the face to which it is strongly adsorbed. Therefore it appears that the desolvation of the crystal surface is the rate-determining step for the speed of growth of the face. It is important to note that the "tailored" solvent occupies a natural surface site, its binding is not weakened by completion of the surface layer structure. We may extend this argument to "normal" solvents which are strongly bound to the surface. But if the solvent is strongly bound at a subset of sites and repelled (or very weakly adsorbed) at the remaining surface sites (pockets for example) a different set of rules may apply according to the results on γ-glycine and (R,S)-alanine [2,3]; the surface may be exposed to a cycle of solvent binding at a subset of surface sites, solute adsorption at the "free sites", followed by solvent expulsion, leading to relative fast growth of this face by a kind of "relay" mechanism. This "relay" mechanism is not confined to crystals with polar axes, but should be general. Perhaps we may also conclude that, in general, the inhibiting properties of the solvent cannot be estimated by simply evaluating the average binding energy of solvent to a specific face, but one must take into account its relative binding properties at the different surface sites. Moreover we are of the opinion that the fast growth arising from strong adsorption of solvent on a subset of sites and repulsion on the others might be the connecting link to surface roughening arising from strong solvent-surface interactions.

Acknowledgements

We thank the US-Israel Binational Foundation and the Minerva Foundation for financial support.

4.   REFERENCES AND NOTES

1.   This manuscript contains sections taken from a manuscript recently published in J. Am. Chem. Soc. (1990), **112**, 6215; these include a description of the the crystal growth of two crystalline hydrates (asparagine·H_2O and rhamnose·H_2O) and the growth and dissolution of (R,S)-alanine and γ-glycine in H_2O, CH_3OH, CH_3CH_2OH.

2.
(a) Bennema, P.; Gilmer, G.; In "Crystal Growth; An Introduction", Hartman, P., Ed.; North Holland:Amsterdam, 1973, p. 274
(b) Elwenspoek, M.; Bennema, P.; van der Eerden, J.P.; *J. Cryst. Growth*; (*1987*), **83**, 297.
(c) Bennema, P.; van der Eerden, J.P. In "Morphology of crystals" Terra Scientific Publishing Company, Tokyo 1987, p. 1-75

3.   Bourne, J. R.; Davey, R. J.; *J. Cryst. Growth*; (*1976*), **36,** 278.
     Bourne, J. R.; Davey, R. J.; *J. Cryst. Growth*; (*1976*), **36,** 287.

4.
  (a) Wells, A.F., *Phil. Mag.*; (*1946*), **37,** 184.
  (b) Wells, A.F. *Discuss. Faraday Soc*; (*1949*), **5,** 197.
5.   Wireko, F.C.; Shimon, L.J.W.; Berkovitch-Yellin, Z.; Lahav, M.; Leiserowitz, L.; *J. Phys. Chem.*; (*1987*), **91,** 471

6.   Berkovitch-Yellin, Z. *J. Am. Chem. Soc.*; (*1985*), **107,** 8239

7.   Davey, R.J. *J. of Cryst. Growth*; (*1986*), **76,** 637.
     R.J. Davey, B. Milisavljevic, and J.R. Bourne, *J. Phys. Chem*; (*1988*), **92,** 2032.

8.   Addadi, L.; Berkovitch-Yellin, Z.; Weissbuch, I.; van Mil, J.; Shimon, L.J.W.; Lahav, M.; Leiserowitz, L.; *Angew. Chem. Int. Ed. Engl.*; (*1985*), **24,** 466

9.   Addadi, L.; Berkovitch-Yellin, Z.; Weissbuch, I.; Lahav M.; Leiserowitz, L.; In "Topics in Stereochemistry"; Wiley and Sons:(1986), *vol. 16*, p. 1-85.

10. Berkovitch-Yellin, Z.; Addadi, L; Idelson, M; Leiserowitz, L; Lahav, M; *Nature (London)*; (*1982*), **296,** 27.

11. Jonsson P.G; *Acta Cryst.* (*1971*), **B27,** 893 and references cited therein.

12.
  (a) Taylor R.; Kennard O.; *J. Am. Chem. Soc.*; (*1982*), **104;** 5063.
  (b) Berkovitch-Yellin Z.; Leiserowitz L.; *Acta Cryst,* (*1984*), **B40,** 159.

13. Verbist, J. J.; Lehman, M. S.; Koetzle, T. F.; Hamilton, W. C.; *Acta Cryst.*; (*1972*), **B28,** 3006

14. Addadi L.; Berkovitch-Yellin, Z.; Domb, N; Gati, E.; Lahav, M.; Leiserowitz, L; *Nature*; (*1982*), **296,** 21

15. Indeed asparagine monohydrate when grown in the presence of aspartic acid crystallizes as {010} plates. Low temperature neutron diffraction studies of the mixed crystal of asn/asp demonstrated that the aspartic acid is preferentially absorbed on the {010} face, inhibiting growth along the *b* axis [16b].

16.
  (a) Wang, J. L.; Berkovitch-Yellin, Z.; Leiserowitz, L.; *Acta Cryst.*; (*1985*), **B41,** 341.
  (b) Weisinger-Lewin, Y.; Frolow, F.; McMullan, R.K.; Koetzle, T .F.; Lahav, M. and Leiserowitz, L.; *J. Am. Chem. Soc.*; (*1989*), **111,** 1035.

17. Takagi, S.; Jeffrey, G.A.; *Acta Cryst.*; (*1978*), **B34,** 2551

18. The symbol {*hkl*} designates all the symmetry related faces, (*hkl*) designates the face specified. Thus for spacegroup $P12_11$, with *b* the polar axis, {*hkl*} specifies (*hkl*) and ($\bar{h}k\bar{l}$).

19.
- (a) Wright, D.A; Marsh, R.E; *Acta Cryst.*; (*1961*), **15**, 54.
- (b) Koetzle, T.F;Lehmann, M; Verbist, J.J; Hamilton, W.C; *Acta Cryst.*; (*1972*), **B28**, 3207.

20.
- (a) Iitaka, Y.; *Acta Crystallogr*; (*1961*), **14**, 1.
- (b) Kvick, A.; Canning, W. M.; Koetzle, T. F.; Williams, G. J. B.; *Acta Cryst.*; (*1980*), **B36**, 115

21.　　Donohue, J.; *J. Am. Chem. Soc.*; (*1950*), **72**, 949

22.
- (a) Shimon, L.J.W.; Wireko, F.C.; Wolf, J.; Weissbuch, I.; Addadi, L.; Berkovich-Yellin, Z.; Lahav, M.; Leiserowitz, L.; *Mol. Cryst. Liq. Cryst.*; (*1986*); **137**, 67
- (b) Shimon, L.J.W.; Lahav, M.;Leiserowitz, L.; *J. Am. Chem. Soc.*; (*1985*), **107**, 3375.
- (c) The crystals of (*R,S*)-alanine and γ-glycine exhibit polar axes and so the absolute arrangement of the molecules with respect to the polar *c* axis had to be determined.[17a,17b].

23.　　This was made manifest by distorted hydrogen bond geometry in crystal structures [24,25], by crystal growth[26] and etching experiments[27], from selective occlusion of "tailor-made" acid additives in host amide crystals[16,25], and by atom-atom potential energy calculations.[26]

24.　　Huang, C.; Leiserowitz L.; Schmidt G. M. J., *J. Chem. Soc., Perkin Trans. 2*; (*1973*), 503

25.　　Vaida M.; Shimon L.J.W.; van Mil J.; Ernst-Cabrera K.; Addadi L.; Leiserowitz L.; M. Lahav; *J. Am. Chem. Soc.*; (*1989*), **111**, 1029

26.　　Berkovitch-Yellin Z.; van Mil J.; Addadi L.; Idelson M.; Lahav M.; Leiserowitz L.; *J. Am. Chem. Soc.*; (*1985*), **107**, 3111

27.　　Shimon, L.J.W.; Lahav, M.; Leiserowitz, L.; *Nouv. J. Chemie*; (*1986*), **10**, 723

28.　　The water molecules bound to the outermost layer of CO_2^- groups via O-H-O hydrogen bonds provides additional stabilization energy to the (001) surface layer which has a low molecular density with all molecular dipoles pointing approximately in the same direction.

29.　　Similar evidence for inhibition at both ends of the polar crystals is provided by preliminary crystal growth experiments of γ-glycine in the presence of ionic species such as $Ca(OH)_2$, NH_4Cl, LiCl, NaCl, KCl, RbCl and CsCl, which indicate inhibition at both ends of the polar axis, restricting growth to within the *ab* crystal plane.

30.　　O. Butbul, M.Sc. thesis, Weizmann Institute of Science, Rehovot, Israel, (*1986*).

7

Intermolecular interactions in the crystals of simple cyclic β-diketoalkanes

Andrzej Katrusiak

Introduction

The crystal structures of several simple cyclic β-diketoalkanes have been studied at high pressures; they include 1,3-cyclopentanedione (hereinafter CPD), 1,3-cyclohexanedione (CHD) and 2-methyl-1,3--cyclopentanedione (MCPD). In the crystalline state the molecules of these compounds are present in enol form and are linked by short (Table 1) asymmetric and nearly linear hydrogen bonds. Several features are common to these structures:
- the molecules are hydrogen-bonded into chains,
- the conjugated π-electron bond system and the enolic hydrogen atom (O=C-C=C-OH) as well as the adjacent atoms form a planar fragment of the molecules,
- the molecules in one chain are coplanar,
- the chains of the molecules are arranged in the crystals in such a way that each chain has one neighbouring chain on each side forming sheets of chains. The interactions beween the chains within one sheet are dominated by electrostatic forces between net atomic charges of atoms in the conjugated-bond systems of the molecules. This can be deduced from the close distances between these fragments of molecules of neighbouring chains. Figure 1 shows net atomic charges of the CPD molecule; the structure of the CPD crystal is presented in Figure 2.
- in the crystals of CHD and MCPD the planes of the chains are perpendicular to the planes of the sheets, while in CPD these planes are at 78°,
- the interactions between the sheets can be classified as mainly of van der Waals type.

Despite these similarities between the CPD, CHD and MCPD structures, each of them reacts to high

hydrostatic pressures in a different way:
- the MCPD structure is very stable – the closest molecules of two neighbouring chains do not change their position significantly with respect to one other (Katrusiak, 1990b),
- the crystals of CHD undergo a strong structural transformation at about 0.1 GPa – alternate chains shift by about 0.8 Å with respect to near neighbours, the molecules rotate by *ca.* 10° about the axis perpendicular to their rings and the enolic hydrogen atom interchanges its donor and acceptor sites in the hydrogen bond simultaneously with the change of sequence of double and single bonds in the conjugated–bond fragments. The transformation is accompanied by a strong deformation of the crystal shape, but it does not change the symmetry of the crystal. At high pressures atom C(5), which is disordered at ambient pressure, becomes ordered (Katrusiak, 1990c).
- the CPD crystals break into very small pieces when pressure is elevated to about 0.4 GPa. Although the pieces were very small, it was still possible to observe a significant change of the shape of the fragments which broke off the edges of rectangular crystal samples (Katrusiak, 1990a).

Table 1. Selected crystallographic and structural data for 1,3–cyclohexanedione (CHD), 1,3–cyclopentanedione (CPD) and 2–methyl-1,3–cyclopentanedione (MCPD) at atmospheric pressure. Parameter ρ is the angle between the central line of the chain (which is colinear with the two–fold screw axis in CPD and with the intersection of plane c and the plane of the chain in CHD) and the line drawn through oxygen atoms of the molecule. Parameter δ is defined in the text; l is the interval of chain per molecule.

	CHD	CPD	MCPD
space group	$P2_1/c$	$C2/c$	$C2/m$
a (Å)	6.145(2)	7.451(2)	12.766(2)
b (Å)	11.711(3)	12.853(3)	6.807(2)
c (Å)	8.196(1)	10.754(2)	6.4814(8)
β (°)	99.43(1)	111.90(2)	93.94(1)
V (Å^3)	581.9(1)	956.2(2)	561.9(2)
l (Å)	7.086(5)	6.426(3)	6.4814(8)
ρ (°)	3.45(1)	5.32(1)	–
δ (Å)	1.871(2)	0.0	–
H–bond:			
length (Å)	2.561(4)	2.542(8)	2.598(2)
configuration	*anti–anti*	*anti–anti*	*anti–syn*

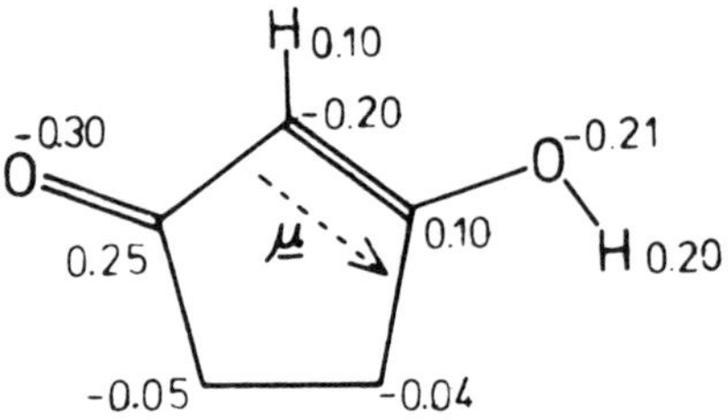

Fig 1. The net atomic charges of the 1,3-cyclopen-tanedione (enol form) molecule calculated by the MNDO method (Dewar & Thiol, 1977) based on the 0.1 MPa structure (Katrusiak, 1989). The H-atoms at C(4) and C(5) have small positive charges smaller than 0.05 e; the direction of dipole moment μ of the molecule is also shown (μ=4.53 Debye).

Because the samples were damaged, it was impossible to analyse the high-pressure struc-cure of CPD by single-crystal X-ray diffraction methods.
The main aim of this study is to analyse the inter-molecular interactions in the crystals of CHD, CPD and MCPD and to suggest possible structural changes which can take place in CPD at elevated pressures.

Chain structure

The structure of chains present in the CPD, CHD and MCPD crystals deserves special attention. The molecules can be connected into chains in various ways, depending on the stereochemical configuration of the hydrogen bonds (Etter, Urbanczyk-Lipkowska, Jahn & Frye, 1986).

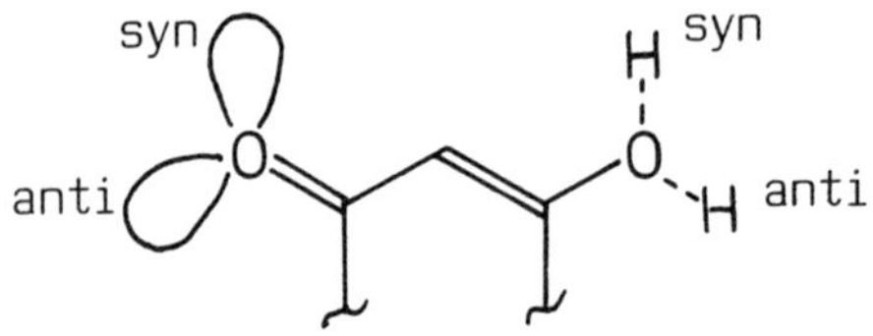

The configuration of the hydrogen bond influences the properties of both the crystal and the hydrogen bond, as it involves two possible sites of the proton. Whereas the H-bonds in CPD and CHD both have the *anti-anti* configuration, that in MCPD is

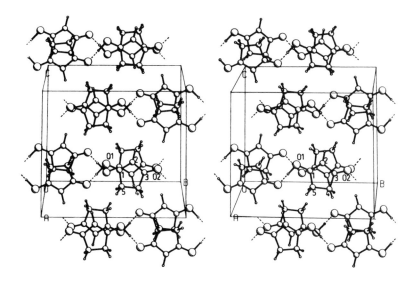

Fig 2. A PLUTO (Motherwell, 1976) stereodiagram of the 1,3-cyclopentanedione (ambient pressure) structure. Hydrogen bonds are represented as broken lines.

anti-syn (configuration of the hydroxyl group is specified first). Thus the two sites of the hydrogen atoms in CPD and CHD are similar with respect to molecules in one chain, but in MCPD are different. Figure 3 presents the arrangement of the CHD and MCPD molecules in the chains. It is characteristic that the dipole moments of the molecules form a small angle with the chain direction. Moreover, the molecules are inclined in a way which is imposed by the geometry of the hydrogen bonds and minimizes the angle between the dipole moments and the direction of the chains. In the chains of CPD and CHD, this inclination of the molecules leads to a larger distance of the hydroxyl oxygen atom from the central line of the chain than the displacement of the carbonyl oxygen atom from this line. This condition is also fulfilled in the high-pressure phase of CHD, when the proton interchanges its position in the hydrogen bond and, consequently, it changes the direction of the dipole moment of the molecule. It appears that the change of orientation of the CHD molecule with respect to the central line of the chain at high pressures [which results from the phase transition driven by

Andrzej Katrusiak

the onset of ordering of atom C(5) in the rings of
the CHD molecules] precedes the change of the site
of the enolic proton and that it is the main cause
of this change (Katrusiak, 1990c).

Arrangement of chains

   The   sheets   of   electrostatically   interacting
chains have a different pattern in CPD, CHD and
MCPD.   In   MCPD   these   sheets   are   built   of
anti-parallel chains; the dipole moments of the MCPD
molecules of two neighbouring chains have opposite
directions.   In the crystals of CHD the polarization
of all the chains within one sheet is the same.   The
chains of CPD, as in CHD crystals, have the same
polarization along chain direction [010] within one
sheet (see Figures 2 and 4).   However, the
significant difference between the structures of CHD
and CPD is that, while in CHD there are no symmetry
elements   which   would   restrict   the   relative
displacements, δ, of the chains along the chain
direction, in CPD there is a two-fold axis along
[010]   relating   the   neighbouring   chains   and
the y-coordinates of the near molecules in the
neighbouring chains are the same (parameter δ, which
describes the displacement of two adjacent molecules
of the neighbouring chains within one sheet along
the chain direction, is restricted to 0.0 Å).

   The   arrangement   of   chains   within   one   sheet
appears to be an important factor which can
stabilize or destabilize the crystal structure at
varied physical conditions, in particular at high
pressures.   The MCPD structure, the chains of which
are *anti*-parallel within one sheet, is exceptionally
stable, while the structures of CPD and CHD — the
chains of which have the same polarization within
the sheets — undergo strong structural transformations.

   It was suggested (Katrusiak, 1990c) that the
initial   force   triggering   the   shifts   of   near
neighbouring chains during the structural transformation of the CHD crystals was a shear strain, which
was generated by electrostatic forces between close
chains.   The   electrostatic   forces   between   the
molecules   can   be   roughly   approximated   by
dipole-dipole interactions.   When the crystal undergoes a structural phase transition the electrostatic
forces can destabilize the structure and generate
the shear strains tending to reduce overlapping of
the close molecules and so to decrease the energy of

interactions between their dipole moments (when the crystal is compressed and the distances between chains get shorter, the interactions between the dipole moments of molecules become stronger). In this way parameter δ in CHD changes from 1.87 Å at ambient pressure to about 2.4 Å at 0.2 GPa and to about 2.7 Å at 1.0 GPa. According to this hypothetical mechanism of the transformation, all other changes observed in the CHD crystal (including the interchange of donor and acceptor sites by the enolic H-atom) at high pressure would be a consequence of the initial ordering of atom C(5) and of the shift of the chains. This model implies that the intermolecular interactions between the sheets of chains would be of minor importance for the occurrence of the transformation. The structural changes at high pressures in CHD are well documented; this structure has been determined at 0.1 MPa (Etter *et al.*, 1986) and at 0.52, 1.19 and 1.90 GPa (Katrusiak, 1990c); most recently this structure was also studied at various temperatures. The structure of MCPD was determined at 0.1 MPa (Katrusiak, 1989) and at 1.50, 2.40 and 3.01 GPa (Katrusiak, 1990b).

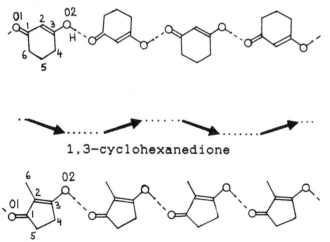

1,3-cyclohexanedione

2-methyl-1,3-cyclopentanedione

Fig. 3. The chains of the CHD and MCPD molecules in the crystal structure and the arrangement of their dipole moments.

Phase transition in CPD

On the basis of the mechanism of structural
transformation presented for CHD, a similar
mechanism can be suggested for the phase transition
of the analogous structure of CPD. At room
temperature and ambient pressure, strong vibrations
of the CPD molecules were observed (Katrusiak,
1990a), which can play a similar role in the transformation of CPD as the disorder of C(5) in the
structural transformation of the CHD crystals. One
can also expect that, when the CPD crystal is
subjected to high pressures, the separation between
the chains is compressed and the interactions
between similarly oriented dipole moments (Figures 1
and 4) somewhat increase the repulsive forces
between the molecules. These repulsive
dipole-dipole interactions could be reduced by
shifts of the chains, which would decrease the
overlap between the close molecules and decrease the
distances between the dipoles. However, unlike in
the CHD structure, in CPD there is no indication
(resulting from the structure symmetry) of the
direction in which such shifts could be made — as
the chains are related by a diad, each of the
directions along [010] is equally probable. The
direction of displacements can be settled by local
defects of the CPD crystal. The shifts, when
effected, break the symmetry of the diad; the
symmetry of the CPD structure is lowered and, due to
large strains on the boundaries of the regions which
assume different directions of the shifts, the
sample crystal breaks into small pieces (Fig. 4b).
By analogy with CHD, it can be assumed that the
high-pressure structure of CPD is monoclinic, but
the new **b'** axis (of the high-pressure phase) is
probably perpendicular to the low-pressure **b** axis —
as suggested by the shape of small fragments of the
damaged CPD samples. Currently, we cannot comment
on the position of the enolic H-atom, except that it
remains ordered in the hydrogen bond and that the
CPD molecules are probably inclined to the central
line of the chain in similar way as was observed in
the low pressure phase of CPD and both phases of
CHD. The CPD crystals used in the high-pressure
experiments were of low quality owing to
difficulties in crystallization of this chemically
unstable substance. It is possible that small and
highly-perfect CPD samples could sustain the
transition and could be pulled into a single
high-pressure-phase crystal.

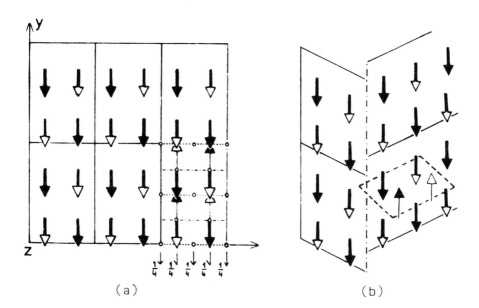

(a)   (b)

Fig 4. (a) The 1,3-cyclopentanedione molecules in one sheet at z=0.25 represented schematically as arrows (full and open arrow-heads correspond to the positive and negative z-components of μ, respectively); six unit cells are included, the bottom-right cell contains all the molecules in the cell (the molecules at z=0.75 are drawn with thin lines) and shows the symmetry elements in this structure; in this projection the dipole moments should have small zigzag-like displacements along [010] which have been neglected in this drawing.
(b) A possible arrangement of the molecules in the high-pressure structure – the crystal is broken in two parts along the dash-and-dot line. The solid lines correspond to the low-pressure unit-cell edges and the broken line indicates a possible unit-cell of this hypothetical high-pressure structure.

Conclusions

The crystal structures of 1,3-cyclopentanedione, 2-methyl-1,3-cyclopentanedione and 1,3-cyclohexane-dione have been compared. The molecules of these compounds (enol forms) are hydrogen-bonded into chains, but several features of the chains and of their arrangement in the crystal are different. These differences appear to cause drastic differences in the properties of these crystals and in their response to high hydrostatic pressures. The mechanism of the structural transformation of the CHD crystals affords an explanation of the pressure—induced phase transition in CPD. This mechanism implies an important role of electrostatic interactions between close neighbouring chains for the structural transformation which takes place when the crystal undergoes a phase transition. It is suggested that, in the structures where the dipole moments of close molecules have a similar orientation, the molecules tend to reduce their overlapping when their separation is compressed at high pressures. This assumption seems to be strongly supported by the absence of the significant structural changes in the crystals of MCPD, where the dipole moments of the molecules in adjacent chains have *anti*—parallel orientation. The proposed mechanism of the structural transformation in the crystals of simple cyclic β—diketoalkanes indicates an important role of the electrostatic interactions for the structure and properties of such molecular crystals. The hypothetical mechanism of the pressure—induced transformation of CPD presented in this paper implies that CPD is a ferroelastic crystal undergoing an exceptionally strong phase transition.

Acknowledgement

This study was partly supported by the Polish Academy of Sciences, Project CPBP 01.12.

References

Dewar M.J.S. & Thiel W. (1977). *J. Am. Chem. Soc.* **99**, 4899–4907.

Etter M.C., Urbanczyk-Lipkowska Z., Jahn D.A. & Frye J.S. (1986). *J. Am. Chem. Soc.* **108**, 5871–5876.

Katrusiak A. (1989). *Acta Cryst.* **C45**, 1897–1899.

Katrusiak A. (1990a). *Acta Cryst.* **C46**. In the press.

Katrusiak A. (1990b). *High-Pressure Research.* Submitted.

Katrusiak A. (1990c). *Acta Cryst.* **B46**, 246–256.

Motherwell W.D. (1976). PLUTO. *Program for plotting molecular and crystal structures.* University of Cambridge, England.

8
Strain in cyclo-annulated aromatic systems
R. Boese, D. Bläser, N. Niederprum, and T. Miebach

I. LOW-TEMPERATURE CRYSTALLIZATION TECHNIQUES

One of the essentials for organic crystal structure investigations is the availability of a single crystal. Because of the special problems of investigation and the difficulties in handling crystals of highly strained molecules, which are usually extremely sensitive to heat or are liquid under ambient conditions, the first part of this contribution will describe the techniques for crystallizing samples and for getting them onto the diffractometer without damage.

Crystallization of organic compounds often turns out to be a cumbersome business because optimal parameters of most compounds are completely unknown at the start of the experiment. The technique used most is by cooling a solvent in a refrigerator. But this technique does not take into account one of the fundamentals of crystallization: Simply cooling down below the point of saturation causes almost no effect in the region of the highest crystallization speed.

Only below the temperature of highest crystallization speed, crystal seeds are formed, indicated by the solution becoming cloudy. Now increase of the temperature is recommended so that the formation of crystal seeds is reduced and crystals can grow in the region of the optimum crystallization speed.
However this is difficult to carry out in a refrigerator, so it is preferable to use a cryostat with a precise temperature control. Because observation of the process is necessary, the crystallization should be carried out in external vessels, such as shown in Figure 1 (Boese & Bläser, 1989).

The inner vessel (A) has a connection for flushing with a protective gas and is surrounded with a chamber for the coolant (B) from the cryostat. The outer chamber (C) is a silver coated vacuum chamber with a viewing window, in order to avoid icing which might obscure the inner chamber where the crystallization process takes place.

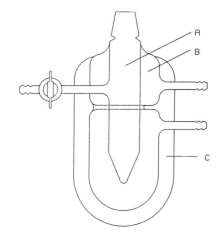

Fig. 1 Vessel for controlled crystallization

Difficulties arise if the crystals are sensitive to room temperature and the selection of crystals must take place under permanent cooling. For this case a selection vessel, such as shown in Figure 2, is well suited. This vessel also consists of three chambers; the inner one (C) is for the selection of the crystals, the next one (B) is for the coolant, and the third one is a vacuum chamber (A). These chambers are covered with a double-walled optical glass (D), glued on the top and allow viewing with a microscope. For the purpose of illumination, the vessel is placed on a light house (E,J). The protective gas is conducted through a glass coil (c) within the cooling chamber to cool down the gas, flushing the inner vessel. There are three inlets (e,d,f) into the inner vessel; two of them are ground glass joints (e,g) for closing them with a

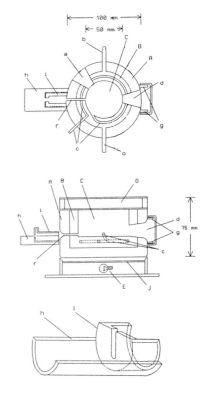

Fig. 2 Selection vessel for low temperatures

glass stopper. Inlet (e) allows connection to the crystallization vessel for transfer of the mother liquid with the crystals; inlet (d), sealed with a rubber septum, allows manipulation of the crystals with a needle, pretruding the rubber seal. Into inlet (f) the capillary for the selected crystal is placed. Outside inlet (f) a semi-cylindrical plastic slide is attached (h) serving as a holder for a small box, cut open on one side. This can be filled with finely powdered dry ice. The slot in the box fixes the metal holder into which the capillary is glued and which is placed onto the goniometer head after withdrawing the capillary into the dry ice and melting off. At the diffractometer, the dry ice is blown off by the cold gas stream of the low-temperature device and the box is removed sidewards.

For samples with very low melting points (less than -50°C, but higher than the lowest temperature provided by the low-temperature device), crystal growing "in situ", i.e. directly on the diffractometer, is recommended (Brodalla, Mootz, Boese, Oßwald, 1985). The most powerful procedure for this operates with focused infrared light, provided by a halogen lamp; see Figure 3.

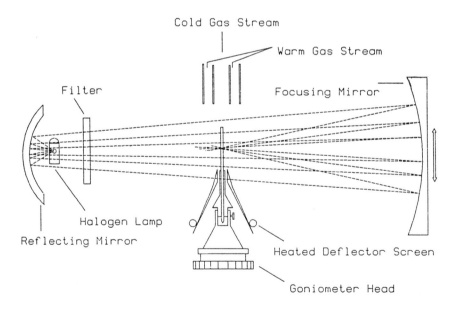

Fig. 3 Radiation crystallisation device on the diffractometer

The light source is a commercially available slide projector with the lenses replaced by a cut-off filter for UV light. On the opposite side a precise ground parabolic mirror is placed, focusing the heat onto the capillary, which is cooled below the melting point of the sample. By moving the parabolic mirror, a miniature zone-melting procedure can be performed by shifting the molten zone along the capillary. This also provides the opportunity to observe the crystallization process with polarized light through the microscope at the diffractometer. Crystal quality may be checked readily with rotation photographs or by peak scanning without removing the crystallization apparatus.

For transferring the samples to the diffractometer, which are gases at room temperature, another simple technique is utilized. The capillary is glued to a vacuum line with its funnel on top (Figure 4).

A metal holder, later placed into the goniometer head, is glued to the capillary and fixed with a clip probe. The vacuum line is evacuated and the gaseous sample condensed into the capillary by cooling with liquid nitrogen.

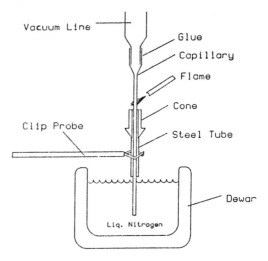

Fig. 4 Procedure to transfer gases into capillaries

After melting-off, a "clothes-peg"-like nozzle, as shown in Figure 5, is attached to the metal holder, to hold the capillary. The nozzle of the "clothes-peg" opens in two halves. Through one leg of the peg a glass tube transfers liquid nitrogen from a styropor box to the capillary. This can be brought to the diffractometer and the peg is then turned by 180° so the capillary stands upright and is placed into the cold gas stream of the low-temperature device at the diffractometer. The peg can be removed now by opening the legs. The capillary is thus cooled all the time.

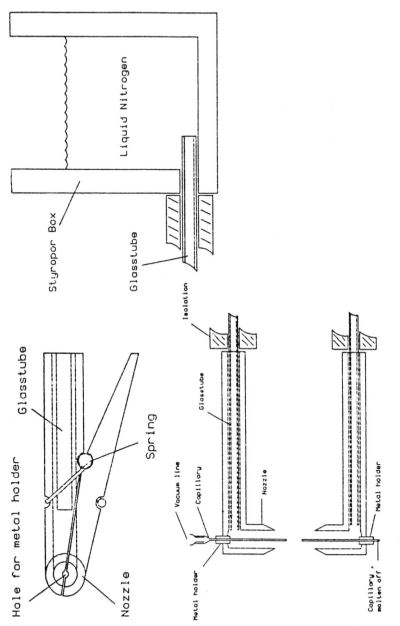

Fig. 5 "Clothes-peg" with incorporated nozzle for transfer of samples at low temperatures

II.   COMPARISON OF MOLECULAR STRUCTURES OF ANNULATED AROMATIC SYSTEMS

Mills and Nixon (1930) suggested that when aliphatic rings are annulated to benzene one of its Kekulé structures should be preferred (Figure
6). Since this effect was expected to increase with a decreasing number

Fig. 6 The Mills-
Nixon effect

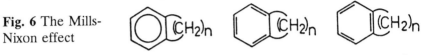

of the annulated aliphatic ring atoms, the effect should be most significant in cyclopropabenzene . The Mills-Nixon-Effect is still being
discussed in its theoretical aspects (Cheung, Cooper, Manat, 1971;
Apeloig & Arad, 1986; Eckert-Maksić, Hodošcek, Kovacek, Mitic,
Maksić, Poljanec, 1990) especially with semiempirical and ab initio calculations. Two recent reviews by B. Halton (1989) and B.E. Billups
(1988) give a comprehensive survey of cycloproparene chemistry.
Although the structures of some derivatives of cyclopropabenzene have
been investigated earlier, it was not clear whether the substituents
have a great influence on the molecular geometry. The structure of the
parent compound (Figure 7) was determined (Neidlein, Christen,
Poignée, Boese, Bläser, Gieren, Ruiz-Pérez, Hübner, 1988) by the zone-
melting procedure described above.

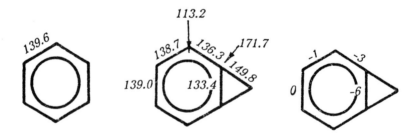

Fig. 7 The structure of cyclopropabenzene, compared to benzene[pm,°]

We found a significant shortening of the bridge bond but no alternation
of bond lengths. Compared to benzene there is a reduction of 6 pm in
the annulated bond, but neighbouring bond lengths are also reduced. An
exocyclic angle of 172° indicates that the system tries to reduce this
angle by shortening of the annulated bond. The comparatively small
innercyclic angle of 113° (7° less than in benzene) and the 120° angles
in the non-annulated side of the ring indicate that the left side of the
ring as drawn remains almost undistorted while the right side suffers all
the strain from the cyclopropane ring.

The situation is comparable to the structures (Boese & Bläser, 1988) of cyclobutabenzene and dicyclobuta[a,d]benzene shown in Figure 8.

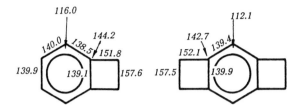

Fig. 8 The structure of cyclobutabenzene and dicyclobuta[a,d]benzene [pm, °]

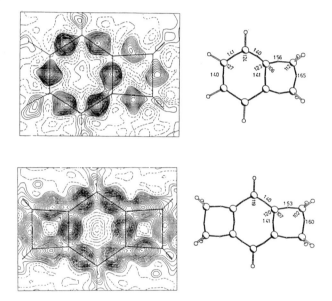

Fig. 9 Difference electron density maps in the molecular planes of cyclobutabenzene (top) and dicyclobuta[a,d]benzene (bottom) with bond lengths and angles based on the bond paths rather than the internuclear lines

By annulation, the aromatic ring system is squeezed at either one or two sides, respectively. Whereas the bond lengths differ only slightly, the innercyclic angles are reduced by 4°(compared to benzene) in the mono-annulated and twice as much (8°) in the doubly-annulated system.

Although electron difference density maps should be viewed critically in general (Boese, 1989), they do provide an explanation for the above results (Figure 9). The electron density maxima are shifted exocyclic with respect to the cyclobutene rings, resulting in an exocyclic shift at the bonds in the benzene ring adjacent to the annulated bonds. If one takes the 'bond path angles' (via the bonding electron maxima) instead of the internuclear bond angles, a much less distorted situation of the bond angles results.

An additive effect of the double introduction of strain can be observed in the structure of the "mixed" molecule, the highly unstable cyclopropacyclobuta[a,d]benzene (Bläser, Boese, Brett, Rademacher, Schwager, Stanger, Vollhardt, 1989) (Figure 10).

The smallest angle in the benzene ring (109°) is a reduction of 11°, and equals the sum of 4° and 7° for cyclopropabenzene and cyclobutabenzene, respectively.

The bond lengths in this system are expanded with respect either to those in cyclopropabenzene or to those in cyclobutabenzene. This expansion is in response to strain

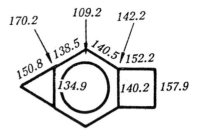

Fig.10 Cyclopropacyclobuta[a,d]-benzene [pm, °]

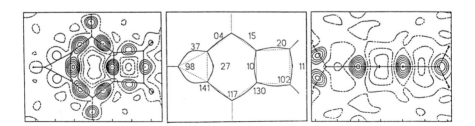

Fig. 11 Difference electron maps of cyclopropacyclobuta[a,d]benzene in the molecular plane (left) and perpendicular in the mirror plane (right). The middle picture gives the distances of the difference maxima (in pm) from the internuclear lines and the bond path angles.

from both sides and is also mirrored in the reactivity of the system. The electron-withdrawing effect of both the cyclopropane and cyclobutane rings should be the reason for the expansion of the aromatic ring system. The difference of strain introduced into the ring from both sides can also be demonstrated by difference density maps (Figure 11).

No alternation of bond lengths could be detected in the benzoannulated small ring compounds.

However, the tendency of an anti-Mills-Nixon-effect in another system could be observed: spiropentane was predicted by ab initio calculations to be bent more easily than twisted if the C2 and C4 atom are bridged (Wiberg, 1985) (Figure 12).

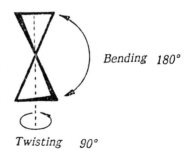

Bending 180°

Twisting   90°

Fig. 12 Bridging spiropentane in the 1,4-position is accompanied either by bending or twisting as shown; in spiropentane the bending and twisting angles are 180° and 90°, respectively

The geometry of spiropentane, determined by X-ray-analysis (Figure 13a) (Boese, Bläser, Gomann, Brinker, 1989), was found to be in approximate agreement with ab initio calculations (Wiberg, 1985; Wiberg, Bader, Lau, 1987). Upon bridging the C2 and C4 atoms of spiropentane with ethane (tricyclo[4.1.0.0^{1,3}]-heptane), the spiropentane is bent more than it is twisted, as predicted by calculation. The resulting cyclopentane ring is puckered and the bond length of the ethano bridge is slightly longer than the bond lengths in cyclopentane (Adams, Geise, Bartell, 1979) (Figure 13b). This indicates that spiropentane works like a tensile spring, pulling the C-C-bond of the ethano bridge apart.

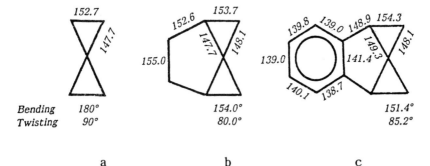

Bending	*180°*	
Twisting	*90°*	

154.0°	151.4°
80.0°	85.2°

a   b   c

Fig. 13 Experimental bond lengths (pm) in spiropentane (**a**), in 'ethano-spiropentane' (**b**), in 'benzospiropentane' (**c**)

This effect is even more pronounced in the benzo bridged spiropentane (Figure 13c), with the resulting cyclopentene ring   also puckered and the bridged C-C-bond of the benzene ring expanded. Whereas twisting of the spiropentane is released, bending is now even stronger than in the 'ethano-bridged' spiropentane. The benzene moiety shows the tendency of bond alternation indicating one Kekulé structure to be predominant as shown in Figure 14.

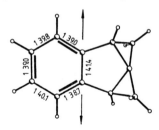

The difference electron density maps demonstrate the bent bonds in spiropentane, becoming asymmetric by bridging in the C2, C4 positions and with the electron density maxima exactly on the internuclear lines at the ethano bridge and in the benzene moiety, respectively (Figure 15).

Fig. 14 The anti-Mills-Nixon-effect of spiropentane, operating like a tensile spring.

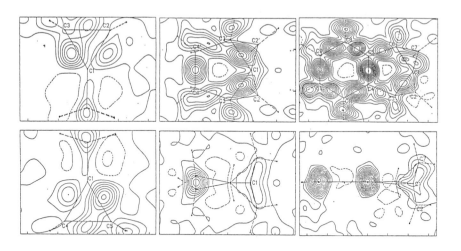

Fig. 15 The difference electron density maps of spiropentane (left), 'ethanospiropentane'(middle) and 'benzospiropentane'(right) in one of the cyclopropane rings (left), in the cyclopentane ring (middle), and in the benzene ring (right)

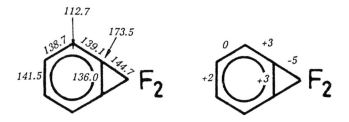

Fig. 16 7,7'-Difluorocyclopropabenzene, bond lengths and angles (left) [pm,°], compared to those of cyclopropabenzene (right)

With electron-withdrawing groups at the cyclopropene ring, for example with fluorine atoms at the previously shown cyclopropabenzene (Figure 16), we find an expansion of the distal bond and a reduction of the vicinal bonds[1]. The effect is transferred also into the opposite bond of the benzene ring.

A similar effect is observed for methylenecyclopropane[2], a gaseous compound at ambient conditions (Figure 17). By comparison with cyclopropane (Nijveld, Voss, Cameron 1982) of the vicinal bond is shortened

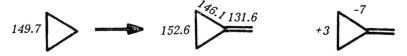

Fig. 17 Comparison of bond lengths (pm) in methylenecyclopropane, and cyclopropane

and the distal bonds are lengthened. These effects can be explained by considering the Walsh orbitals of cyclopropane. Electron-withdrawing groups reduce the size of the neighbouring orbitals; this increases that of the two other orbitals, which are antibonding in the distal bond and bonding in the vicinal bond.

The electron density distribution in 7,7-difluorocyclopropabenzene (Figure 18) shows the expected picture both in the molecular plane and in a perpendicular plane, where the electron density is low between the fluorine atoms and the C atoms because of the electronegativity of the fluorine atoms. The polarisation of the distal bond is indicated by the extraction of electron density from the distal bond via σ-electrons.

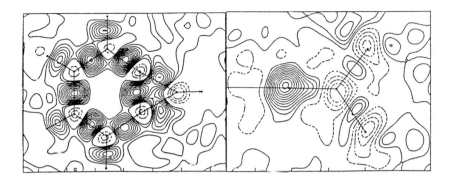

Fig. 18 7,7'-Difluorocyclopropabenzene, difference electron densities in the ring plane (left) and perpendicular through the fluorine atoms (right)

Bond lengths of naphthalene can be well explained by the three resonance pictures (Figure 19), assuming equal contributions of each, as shown. These indicate that three bond lengths should be almost equal

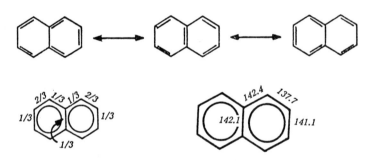

Fig. 19 The three resonance pictures of naphthalene (top), the corresponding double-bond character in its bonds (bottom, left), and the bond lengths [pm] found by X-ray analysis (Brock & Dunitz 1982) (bottom, right)

and one bond should have a higher double-bond character; this is reflected in the experimental bond lengths, found by Brock & Dunitz (1982).

Annulation of naphthalene with cyclopropane shows an almost undis-

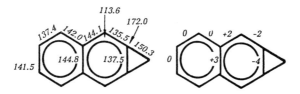

Fig. 20 The bond lengths and angles of cyclopropanaphthalene (left) in comparison to naphthalene (right) [pm, °]

torted left ring but a higher distortion at the annulated side[3] (Figure 20). A small effect was found in the annulated bond when compared to that of cyclopropabenzene, while the central bond of naphthalene is expanded. The concept of bent bonds provides a much better explanation than that given by simple resonance structures. The electron densities indicate a high shift to the maxima exocyclic of the three-membered ring, expanding the central bond in the naphthalene moiety (Figure 21). The annulated bond is reduced, but less than by annulation with benzene, and the effect is transferred much more to the central bond. By double annulation of cyclopropane to naphthalene[4] (Figure 22), bond lengths at both halves are almost the same as in the mono-annulated naph-

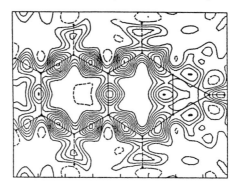

Fig. 21 Difference electron density map in the molecular plane of cyclopropanaphthalene

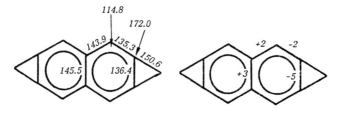

Fig. 22 Bond lengths and angles of dicyclopropa[a,f]naphthalene (left) compared to those in to naphthalene (right) [pm, °]

thalene, except that the annulated bond experiences a greater reduction. The central bond is now expanded because the effect observed for cyclopropanaphthalene is now working from two sides. As shown above, electron-withdrawing groups at the 7-position in cyclopropabenzene reduce the vicinal bonds and expand the distal bond. This can also be observed by annulation of methylenecyclopropene to benzene. In this case the annulated bond is reduced by 15 pm with respect to cyclopropane; whereas for cyclopropabenzene the reduction is only 16 pm. In cyclopropabenzene the vicinal bond is essentially unchanged whereas for the methyline derivative it is reduced in length by 2 pm[5]

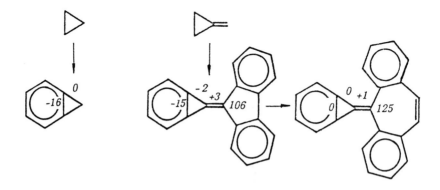

Fig. 23   Comparison of bond lengths [pm] of cyclopropane (left) and methylenecyclopropane derivatives (middle, right) by annulation with benzene; only the differences are given.

(Figure 23, left, middle). The dipole moment of the parent fulvalene has been calculated by ab initio (Apeloig, Karni, Arad 1989) to be in the direction as shown in Figure 24; and the dipole of the dibenzoderivative was measured to be 2.6 D (Halton, Buckland, Mei, Stang 1986; Halton, Buckland, Lu, Mei, Stang 1988). An inverse polarization might be expected for a benzotriaheptafulvalene, in which the positive charge is as a tropylium-cation moiety. This would imply a negative charge in the benzene ring and, as a consequence, a reduction of bond lengths. However, when compared to fulvalene (Figure 23, middle), bond lengths are essentially unchanged[6]. Actually the parent cycloheptatrienylidene was calculated to have a dipole giving positive charge to the benzene ring and negative charge to the cycloheptatrienylidene ring (Apeloig, Karni, Arad 1989), and was suggested to be non-planar as a consequence (Halton 1989). This has been found in the system shown in Figure 23 (right) with torsion angles of 141° from the atoms of the fulvalene double bond to the neighbouring atoms in the cycloheptatrienyl ring. The dipole moment was measured to be 1.2 D (Halton, Buckland, Mei, Stang 1986; Halton, Buckland, Lu, Mei, Stang

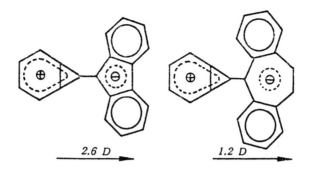

2.6 D   1.2 D

Fig. 24 Suggested direction of dipoles in cyclopentadienyl and cycloheptatrienylmethylenecyclopropabenzene derivatives with experimental values

1988) and should be in the direction as shown on the right side of Figure 24.

Annulation of cyclopropane to naphthalene reduces the annulated bond by 12 pm and expands the vicinal bond by 1 pm (Figure 25, left). With

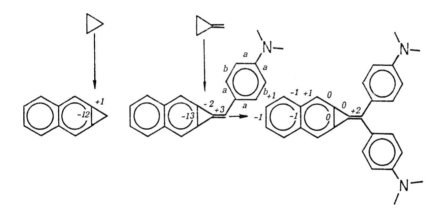

Fig. 25 Comparison of bond lengths of cyclopropane and cyclopropanaphthalene, methylenecyclopropane and the p-N,N dimethylaniline-substituted methylenecyclopropanaphthalene, and its disubstituted derivative

methylenecyclopropane annulated to naphthalene an expansion of the bridge bond might be expected, as was observed for the simpler cyclopropabenzenes. However, with the dimethylanilino group as a substituent the opposite is observed (Halton, Lu, Stang 1988; Halton, Lu, Melhuish 1988) (Figure 25, middle). This indicates that negative charge is shifted into the naphthalene moiety with positive charge at the nitrogen atom. The rings are almost coplanar (naphthalene-phenyl interplanar angle 5.3°)[7]. The deviation from planarity of the nitrogen atom (distance of 28.3 pm from the plane of surrounding atoms), the short N-C(phenyl)-distance (140 pm) and the difference of the mean bonds (a - b)(3 pm) (Figure 25), all support the assumption of dipole direction.

With two p-N,N-dimethylanilino substituents[8], the dipole is higher (Figure 26, right) (Halton, Lu, Stang 1988; Halton, Lu, Melhuish 1988), and this increases the bond length of the fulvalene double bond by 2 pm. The most significant changes in the naphthalene rings are in the left side on going from the mono p-N,N-dimethylanilino to the di-p-N,N-dimethylanilino-substitutedmethylenecyclopropanaphthalene. Although the phenyl rings are twisted against the naphthalene moiety (28.2° and

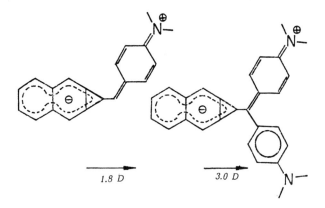

Fig. 26 Experimental dipole moments and resonance structures of p-N,N-dimethylanilinomethylenecyclopropanaphthalene and bis-p-dimethylanilinomethylenecyclopropanaphthalene

27.7°), the deviations from planarity of the nitrogen atoms are 22.9 and 11.9 pm and the bond lengths differences (a - b) are 2.0 and 2.5 pm. Both of these pairs of data, defined as above, coupled with the N-C(phenyl)-distances of 139 and 138 pm provide evidence for the polar structures, one of which is shown in Figure 26 (right).
The interplanar angles of the N-C(methyl)atoms with the phenyl rings

and the deviation from planarity for the p-N,N-dimethylanilino-methylenecyclopropanaphthalene (32.3° and 28.3 pm) and for the bis-p-N,N-dimethylanilino derivative (27.0° and 22.9 pm, 16.7° and 11.9 pm) correlate very well with a linear regression found for the same pairs of values in 106 p-N,N-dimethylanilino derivatives in the Cambridge Crystallographic Data File (Boese 1990).

It has been shown that the concept of bent bonds is a useful qualitative practical tool, able to explain a significant number of bond lengths in highly strained compounds. However, when highly electron-withdrawing groups are present the concept is too simple and more sophisticated models must be applied.

The authors would like to thank Prof. B. Halton, Prof. R. Neidlein, Prof. K.P.C. Vollhardt, Prof. U.H. Brinker for kindly providing the samples for the X-ray investigations. We are gratefully indebted to Prof. B. Halton for helpful discussions.

References

Adams, W.J., Geise, H.J. & Bartell, L.S. (1979) *J. Am. Chem. Soc.*, **92**, 5013.

Apeloig, Y. & Arad, D. (1986) *J. Am. Chem. Soc.*, **108**, 3241.

Apeloig, Y., Karni, M. & Arad, D. (1989) In *Strain and Its Implications in Organic Chemistry*; A. de Meijere, S. Blechert, Eds.; Reidel: Dordrecht, p. 457.

Billups, W.E., Rodin, W.A. & Haley, M.M. (1988) *Tetrahedron*, **44**, 1305.

Bläser, D., Boese, R., Brett, W.A., Rademacher, P., Schwager, H., Stanger, A. & Vollhardt, K.P.C. (1989) *Angew. Chem. Int. Ed. Engl.*, **28**, 206.

Boese, R. (1989) *Chem. i. u. Zeit*, **23**, 77.

Boese, R. (1990), unpublished results.

Boese, R. & Bläser, D. (1988) *Angew. Chem. Int. Ed. Engl.*, **27**, 304.

Boese, R. & Bläser, D. (1989) *J. Appl. Crystallogr.*, **22**, 394.

Boese, R., Bläser, D., Gomann, K. & Brinker, U.H. (1989) *J. Am. Chem. Soc.*, **111**, 1501.

Brock, C.P. & Dunitz, J.D. (1982) *Acta Crystallogr.*, **B38**, 2218.

Brodalla, D., Mootz, D., Boese, R. & Oßwald, W. (1985) *J. Appl. Crystallogr.*, **18**, 316.

Cheung, C.S., Cooper, M.A. & Manat, L.L. (1971) *Tetrahedron*, **27**, 689; 701.

Eckert-Maksić, M., Hodošček, M., Kovaček, D., Mitić, D., Maksić, Z.B. & Poljanec, K. (1990) *J. Mol. Struc.*, **206**, 89.

Halton, B. (1989) *Chem. Rev.*, **89**, 1161.

Halton, B., Buckland, S.J., Mei, Y. & Stang, P.J. (1986) *Tetrahedron Let.*, **27**, 5159.

Halton, B., Buckland, S.J., Lu, Q., Mei, Y. & Stang, P.J. (1988) *J. Org. Chem.*, **53**, 2418.

Halton, B., Lu, Q. & Stang, P.J. (1988) *J. Chem. Soc., Chem. Commun.*, 879.

Halton, B., Lu, Q. & Melhuish, W.H. (1988) Unpublished observations.

Mills, W.H. & Nixon, T.G. (1930) *J. Chem. Soc.*, 2510.

Neidlein, R., Christen, D., Poignée, V., Boese, R., Bläser, D., Gieren, A., Ruiz-Pérez, C. & Hübner, C. (1988) *Angew. Chem. Int. Ed. Engl.*, **27**, 294.

Nijveldt, D., Vos, A. & Cameron, A.F. (1982) *Eur. Cryst. Meeting*, **7**, 153.

Wiberg, K.B. (1985) *J. Org. Chem.*, **50**, 5285.

Wiberg, K.B., Bader, R.F.W. & Lau, C.D.H. (1987) *J. Am. Chem. Soc.*, **109**, 985.

1) Structure data for $C_7H_4F_2$[9]:
Cell dimensions: a = 588.17(8), b = 1279.56(19), c = 1475.72(24) pm, SG: Pbca, Z = 8, T = 110 K, 1639 unique, 1508 observed intensities $(F_0 \geq 4\sigma(F))$, $2\Theta_{max}$ = 60° (Mo - K$_\alpha$), R = 0.043, R$_w$ = 0.044; Esd's of non-hydrogen bond lengths less than 0.1 pm; High angle refinement with calculated hydrogen positions 2536 unique, 1967 observed intensities, $60° \leq 2\Theta \leq 80°$, R = 0.055, R$_w$ = 0.059;

2) Structure data for C_4H_6[9]:
Cell dimensions: a = 516.43(12), b = 1351.6(3), c = 591.31(14) pm, β = 112.23(2)°, SG: $P2_1/c$, Z = 4, T = 126 K, 1113 unique, 974 observed intensities $(F_0 \geq 4\sigma(F))$, $2\Theta_{max}$ = 60° (Mo - K$_\alpha$), R = 0.045, R$_w$ = 0.048, Esd's of C-C-bond lengths less than 0.1 pm;

3) Structure data for $C_{11}H_8$[9]:
Cell dimensions: a = 1118.4(5), b = 1029.6(5), c = 637.5(2) pm, SG: Pnma, Z = 4, T = 120 K, 2375 unique, 1479 observed intensities $(F_0 \geq 4\sigma(F))$, $2\Theta_{max}$ = 80° (Mo - K$_\alpha$), R = 0.061, R$_w$ = 0.056, Esd's of C-C-bond lengths 0.1-0.2 pm;

4) Structure data for $C_{12}H_8$[9]:
Cell dimensions: a = 571.44(13), b = 631.37(14), c = 1059.29(25) pm, β = 95.745(19)°, SG: $P2_1/n$, Z = 2, T = 117 K, 673 unique, 580 observed intensities $(F_o \geq 4\sigma(F))$, $2\Theta_{max}$ = 50° (Mo - K_α), R = 0.050, R_w = 0.050; Esd's of C-C-bond lengths 0.2-0.3 pm;

5) Structure data for $C_{20}H_{12}$[9]:
Cell dimensions: a = 3182.5(8), b = 564.2(2), c = 1467.8(4) pm, β = 94.20(2)°, SG: $C2/c$, Z = 8, T = 125 K, 1667 unique, 1182 observed intensities $(F_o \geq 4\sigma(F))$, $2\Theta_{max}$= 45° (Mo - K_α), R = 0.059, R_w = 0.062; Esd's of C-C-bond lengths 0.4-0.6 pm;

6) Structure data for $C_{22}H_{14}$[9]:
Cell dimensions: a = 766.11(8), b = 1835.4(3), c = 2103.9(4) pm, β = 90.193(11)°, SG: $P2_1/c$. Z = 8, T = 125 K, 6777 unique, 5098 observed intensities $(F_o \geq 4\sigma(F))$, $2\Theta_{max}$= 55° (Mo - K_α), R = 0.044, R_w = 0.053, two independent molecules, Esd's of mean C-C-bond lengths 0.1-0.2 pm;

7) Structure data for $C_{20}H_{17}N$[9]:
Cell dimensions: a = 627.43(7), b = 746.94(8), c = 3094.5(4) pm, SG: $P2_12_12_1$, Z = 4, T = 120 K, 2588 unique, 2336 observed intensities $(F_o \geq 4\sigma(F))$, $2\Theta_{max}$= 50° (Mo - K_α), R = 0.035, R_w = 0.039; Esd's of non-hydrogen bond lengths 0.3 pm;

8) Structure data for $C_{28}H_{26}N_2$[9]:
Cell dimensions: a = 2899.3(8), b = 1151.4(4), c = 640.4(2) pm, SG: $Pna2_1$, Z = 4, T = 125 K, 2408 unique, 2250 observed intensities $(F_o \geq 4\sigma(F))$, $2\Theta_{max}$= 45° (Mo - K_α), R = 0.029, R_w = 0.032; Esd's of non-hydrogen bond lengths 0.3-0.4 pm;

9) Further details of the crystal structure analyses can be obtained from the Fachinformationszentrum Karlsruhe, Gesellschaft für wissenschaftlich-technische Information GmbH, D-7514 Eggenstein-Leopoldshafen 2, on quoting the description number CSD-320156 (No 1); CSD-320154 (No 2); CSD-320158 (No 3); CSD-320155 (No 4); CSD-320157 (No 5); CSD-320161 (No 6); CSD-320160 (No 7); CSD-320159 (No 8), the author and the monograph citation.

9
Stereochemistry of methyl and methoxy substituents in polycyclic hydrocarbons

C. E. Briant, O. Johnson, D. W. Jones, and J. D. Shaw

The molecular sites of methyl, methoxy and other simple substituents can have a marked influence on the biological activity of polycyclic aromatic hydrocarbons (PAH). Thus the carcinogenic activities of the monomethyl-(MBA), dimethyl-(DMBA) and polymethylbenz[a]-anthracenes range from the negligible in 2-MBA or slight in 1,12-DMBA to the extremely potent in 7-MBA or 7,12-DMBA, even though the structure differences are relatively small. Much the same applies to substitution in phenanthrenes (Johnson et al., 1989), benzfluoranthenes and other PAH.

Whereas MBA and DMBA that have methyl substituents away from the bay region are nearly planar or slightly bowed (type I) (Briant et al., 1985), those with methyl substituents at one or both of the 1 and 12 bay sites suffer appreciable buckling of the carbon framework (type II). The methyl carbons in crystal structures of type-II MBA and DMBA can lie 1Å or more out of the mean molecular plane compared with only 0.1Å for type-I structures, although these usually show slight bowing or other perceptible deviation from planarity. Hydrogen atoms of the substituents are not always located accurately in X-ray analyses of these molecules but approximate positions have been determined for many PAH.

We discuss here the dimensions, orientations and inclinations of methyl and methoxy substituents, mostly in structures of some methylphen-anthrenes, methylbenz[a]anthracenes, and methoxybenzo[j]fluor-anthenes, but also including some less simple PAH structures from data extracted from the Cambridge Crystallographic Data base (Allen et al., 1979).

DIMENSIONS AND ORIENTATIONS OF METHOXY GROUPS

Structures of comparatively few simple PAH with methoxy substituents have been determined. The two C-O bond lengths of methoxy

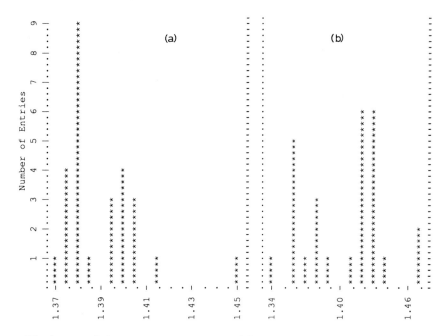

Fig.1   Schematic histogram of (a) C_{aro}-O and (b) O-C_{Me} bond
lengths (Å) in methoxy-substituted PAH.

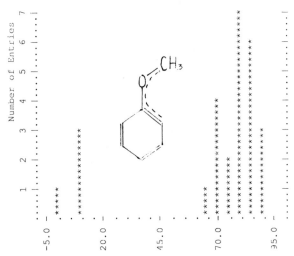

Fig. 2   Schematic histogram of CH_3-O-C_{aro}-C_{aro} torsion angles
(degrees) for aromatic methyl substituents.

substituents in these compounds are usually larger and smaller than 1.40Å, respectively: CH_3-O = 1.40-1.44, centred on 1.42Å , and C_{aro}-O = 1.36-1.40 centred on 1.37Å (Fig. 1). The normal C-O-C angular range is 115-121° with a mean of 117.5°. Although 9-methoxyanthracene (Bart & Schmidt 1971), which has its C-O-C plane perpendicular to the aromatic rings, apparently has both a small C-O-C angle of 112(2)° and an unusually long CH_3-O bond of 1.47(2)Å, its esd's are quite large.

Torsion angles C_{aro}-C_{aro}-O-CH_3 in methoxy-substituted PAH are generally either around 5° or less, corresponding to the methoxy group lying in the ring plane, or else 70-90°, corresponding to the methoxy group sharply out of the plane (Fig. 2). Apparent exceptions, with intermediate torsion angles, are two methoxy groups in a 2,3,7,8-tetramethoxythianthrene complex (Hinriches et al., 1982) but all four are planar in the uncomplexed methoxythianthrene (Hinriches & Klar, 1982) which has O-C-C_{endo} = 115.2° and O-C-C_{exo} = 125.2°. 5-Acetoxy-3,4-dimethoxy-9,10-dihydrophenanthrene (Matsuo et al., 1985) is another sterically crowded case. The 4-methoxy group, confronted across the bay by the 5-acetoxy, is almost perpendicular to the ring plane (with C-O-C = 113.2°) and the buttressing 3-methoxy takes up an intermediate C-O-C-C torsion angle of 60° (C-O-C = 116.9°); endo and exo O-C-C angles are 116.2(2) and 123.8(2)°.

In 7-methoxymethyl,12-methylbenz[a]anthracene (7-CH_2OMe,12-MBA), where there is a methylene group between the PAH and the methoxy group, the C-O-C reverts to a tetrahedral angle of 109.7(3)° (Gilmore & Rae, 1984) and is almost perpendicular to the mean polynuclear plane; the angles of the substituent C_{aro}-CH_2 bond with the two aromatic bonds at the junction differ by 3°.

More typical are two dimethoxybenzo[j]fluoranthrenes (DMBF). In 5,10-DMBF (Briant, et al 1984) carbon and oxygen atoms of the four well-separated methoxy groups of the two independent molecules all lie very close (torsion angles 1-6°) to the mean molecular plane; only one methoxy carbon is as much as 0.2Å out of the plane. The methoxy group of 5-methoxy,7-MBA (Jones & Shaw 1989), without a neighbouring substituent, also lies in the molecular plane. In this compound and in 5,10-DMBF, pairs of methyl hydrogens point above and below the nearest peri hydrogen, H(9), while the C_{aro}-O bond makes dissimilar angles of 124-126°C (on the peri hydrogen side) and 113-114° with the aromatic bonds at the substitution point.

In 7,8-DMBF (Briant & Jones, 1987), on the other hand, the ortho methoxy groups are bent well out of the molecular plane (torsion angles 73 and 83°). The methyl carbon atoms are displaced 1.1 and 1.2Å above and below the plane of their aromatic ring and, in projection, single methyl hydrogens appear to point towards eath other. Angles between the C_{aro}-O bond and the ring bonds at the substitution point

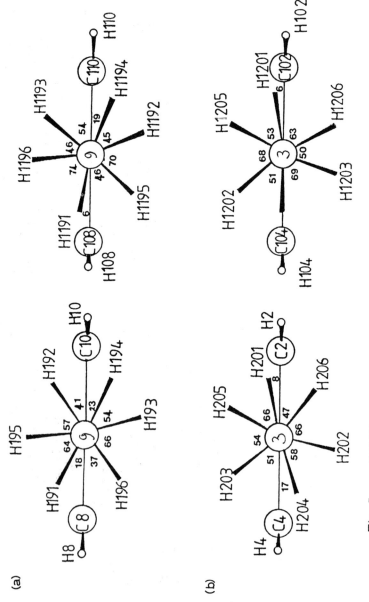

Fig. 3.   3,9-Dimethylbenz[a]anthracene: projection down C_{Me}–C_{aro} bonds showing favoured orientations of methyl hydrogens in the two independent molecules, I and II.

differ by only 2° (121° inside and 119° outside).

For the methoxy aromatics in general C_{aro}-O tends to be rather shorter (around 1.37Å or less) for the planar methoxys than the non-planars (1.38Å or more) and C-O-C tends to be slightly bigger (e.g. 117-118° in 5,10-DMBF and 7-MeOCH$_2$,12-MBA) in planar arrangements than in non-planar (e.g. 113-115° in 7,8-DMBF), in which the oxygen is closer to the neighbouring peri aromatic hydrogen.

In an earlier survey, Anderson et al (1979) had noted that the great majority of unhindered orthodimethoxybenzene derivatives exist in non-planar conformations in the gas phase (contrast meta- and para-dimethoxybenzene) but are coplanar with the ring in their crystal structures. This preference for the planar conformation, unless it is prevented sterically by neighbouring atoms or groups, as by H(6) and H(9) in 7,8-DMBF, has been attributed to the consequent favourable packing in molecular sheets.

7-Hydroxy-2,3,4,6-tetramethoxy-phenanthrene (Pettit et al., 1988) incorporates three adjacent methoxy groups next to a bay region. Much as in the corresponding 9,10-dihydrophenanthrene, the outer 2-methoxy group (like the 6-methoxy) lies in the ring plane but the C-O-C- planes of the inner pair of 3,4-methoxy groups are almost perpendicular to the ring plane. Extreme O-C_{aro}-C_{aro} bond angles are 113.4 - 126.2(4)° in the phenanthrene. A peri hydrogen of a naphthalene moiety is sufficient to cause the 8-methoxy in 3,7,8-trimethoxy-1,4-phenanthrenequinone (Schmalle, et al., 1986) to be twisted well out of the aromatic plane, leaving the adjacent 7-methoxy (and also the 3-methoxy in the quinone ring) in the plane. Similarly, in 2,7-di-0-acetyl-3,4-dimethyloxyphenanthrene (Stermitt et al., 1983), the 2,3-substituents sandwiched between another substituent and a hydrogen across the bay are twisted markedly out of the phenanthrene plane.

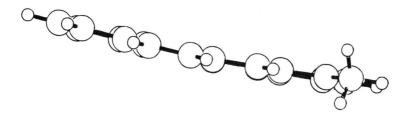

Fig.4 .   8-Methylbenz[a]anthracene: projection down C(13)-C(5) of molecule 1 showing slight curvature of one of the two independent molecules and orientation of methyl group.

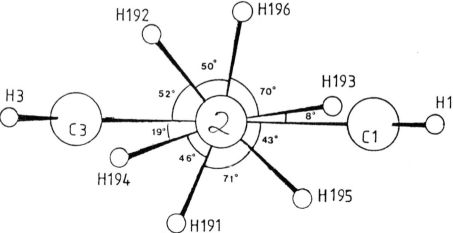

Fig.5.   2-Methylbenz[a]anthracene: in this projection down C_{Me}-C_{aro}, the H(191, 192, 193) methyl orientation is preferred over the H(194, 195, 196).

METHYL GROUP ORIENTATIONS

X-ray refinements suggest that for the methyl groups in 2-MBA (Briant & Jones, 1985) and in the two independent molecules in 3,9-DMBA (Briant, et al., 1986), both fairly planar molecules, the hydrogens occupy multiple rotational sites, i.e. there is more than one favoured orientation.   Fig. 3 shows these for the two methyl rotamers in each of the two independent molecules of 3,9-DMBA, and Fig. 5 the corresponding projection for 2-MBA.

In 6-, 8- and 11-MBA, type-I structures in which the methyls are at non-bay sites in the BCD anthracenic moeity, the methyl groups are all orientated with a pair of hydrogen atoms directed above and below the neighbouring peri aromatic hydrogen (H(7) in 6- and 8-MBA, H(12) in 11-MBA); the third hydrogen lies in the molecular plane pointing away from the bulk of the molecule (Fig.4). Analogously, in the two independent molecules of 5-MBA, methyl hydrogens are directed above and below the peri-naphthalenic H(4), with the third methyl in the same molecular plane as H(6).   A 7-methyl group (as in 5-MeO, 7-MBA or 7,12-DMBA), faced by two peri sites, tends to have hydrogens pointing equally towards H(6) and H(8).   In non-planar or type-II structures with a 12-methyl (bay) substituent, as in 1,12-(Jones & Shaw, 1987) and 7,12-DMBA (Klein, et al., 1987), 7-MeOCH$_2$, 12-MBA and (effectively) 3,6-dimethylchloranthrene (Jones & Shaw, 1989a), the methyl hydrogens of the 12-substituent have a near symmetrical appearance in projection perpendicular to the molecular plane, i.e. there is one C_{aro}-C-H plane roughly perpendicular to the molecular plane (Fig. 6).   However, in these structures, the bay-methyl carbon lies about 1Å out of the PAH plane.

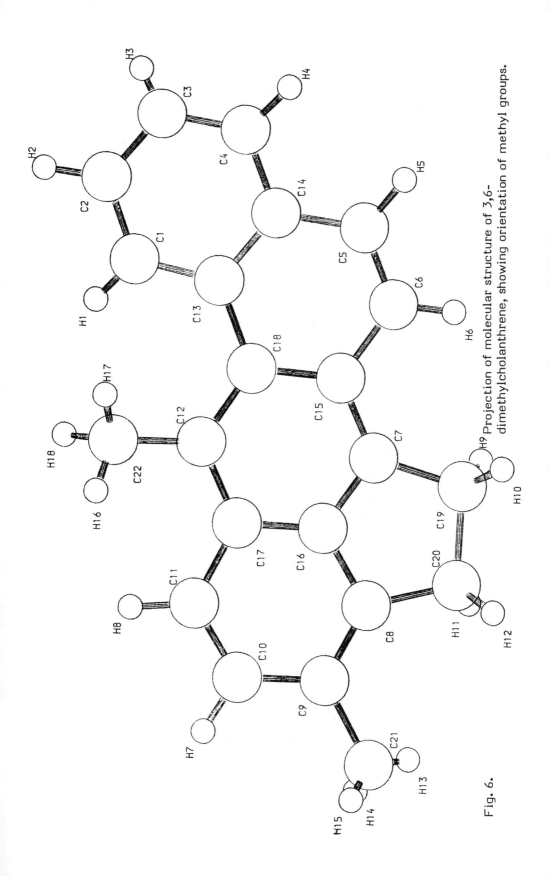

Fig. 6. Projection of molecular structure of 3,6-dimethylcholanthrene, showing orientation of methyl groups.

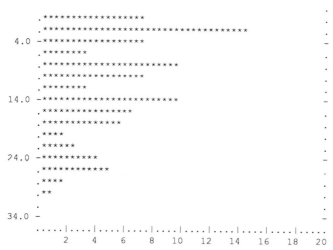

Fig. 7.   Schematic histogram of minimum torsion angles (δ) C_{aro}-C_{aro}-C-H (degrees) for aromatic methyl substituents.

Overall among ordered methylated PAH structures, there is a wide range of minimum torsion angles C_{aro}-C_{aro}-C-H between methyl C-H and ring edge (Fig. 7). Leaving aside apparently long bonds in disordered compounds such as methylanthracenes (e.g. Welberry, et al., 1983), C_{aro}-CH_3 bond lengths in PAH are very closely centred on 1.51Å. There may be a very slight tendency for methyls with one hydrogen virtually in the aromatic plane to be marginally shorter. In 9,10-dimethylphenanthrene (9,10-DMP) (Johnson & Jones, 1989), the two ortho methyl carbons at the K-region are in the molecular plane (total separation perpendicular to the plane of only 0.1Å) with C_{aro}-CH_3 1.498(5) and 1.510(5)Å long. Inner CH_3-C-C bonds of 121-122° slightly exceed outside ones of 118° and single methyl hydrogens point towards one another almost in the molecular plane, i.e. the other pairs of hydrogens point towards the aromatic H(1) and H(8). For the sterically crowded 4,5-DMP, C-CH_3 bond length measurements are available from four independent recent analyses: 1.489 (Cosmo et al, 1987), 1.501 (Armstrong, et al., 1987), 1.502 (Davis 1981), and 1.$\overline{514}$Å (Imashiro et al., 1987). The smallest torsion angles H-C_{meth}-C_{aro}-C_{aro} for the two substituents are 1-4°, i.e. with a C-H in the plane of its ring (although the outer rings have a mutual inclination of 28°) and 26-40°. In 3,4,5,6-tetramethylphenanthrene (Armstrong et al., 1987), the outer ring planes are at 29° to each other but, within each ring, pairs of hydrogen atoms (on 3,4-dimethyls and 5,6-dimethyls) face each other. The methyl carbons each protrude from (above and below) the molecular plane by over 1Å, as in 2,4,5,7-trimethylphenanthrene (Jones & Shaw 1989b).

SUMMARY

Unless prevented by steric interaction, methoxy groups — substituents with lone pairs — in crystals of PAH have preferred orientations making very small angles with the molecular plane of the aromatic rings to which they are attached. This can apply to individual methoxy substituents on separate aromatic rings of different orientations within a molecule as in 9-methoxy-6-(4-methoxybenzyl)-5-(4-methoxyphenyl)benzo[a]xanthene (Ferguson et al., 1987). In such planar methoxy groups, the inner and outer C_{aro}-C_{aro}-O angles can differ by 10° and the C-O-C angles are usually enlarged to 117°. Few methoxy groups take up intermediate torsion angles but pairs of ortho methoxys may have their planes almost perpendicular to the aromatic ring.

Methyl-aromatic bonds in PAH are nearly all very close to 1.51_0 Å. Where possible in nearly planar PAH, the methyl rotamer with two hydrogens bracketing a nearby aromatic hydrogen is favoured. Methyls on adjacent carbons of a ring appear to prefer an orientation with pairs of hydrogens facing each other.

We thank the Yorkshire Cancer Research Campaign for financial support.

REFERENCES

F.H. Allen, S.A. Bellard, M.D. Brice, B.A. Cartwright, A. Doubleday, H. Higgs, T. Hummelink, B.G.Hummelink-Peters, O. Kennard, W.D.S. Motherwell, J.R. Rodgers , D.G. Watson (1979). Acta Crystallogr. B35, 2331-2339.
G.M. Anderson III, P.A. Kollman, L.N. Domelsmith, K.N. Honk (1979). J.Amer.Chem.Soc. 101, 2344-2352
R.N. Armstrong, H.L. Ammon, J.N. Darnow (1987). J.Amer.Chem.Soc. 109, 2077-
J.C.J. Bart, G.M.J. Schmidt (1971). Israel J.Chem. 9, 429-448
C.E. Briant, R.L. Edwards, D.W. Jones and W.S. McDonald (1984). Carcinogenesis 5, 1041-1045
C.E. Briant, D.W. Jones (1985). Cancer Biochem.Biophys. 8, 129-
C.E. Briant, D.W. Jones (1987). Acta Cryst. C 43, 775-778
C.E. Briant, D.W. Jones, R.G. Hazell (1986). Acta Cryst. C 42, 826-829
C.E. Briant, D.W. Jones, J.D. Shaw (1985). J.Mol.Structure, 130, 167-176
R. Cosmo, T.W. Hambley, S. Sternhell (1987). J.Org.Chem. 52, 3119-3539

R.E. Davis (1981). Personal communication via Dr. R. Usha
G. Ferguson, B.L. Ruhl, W.B. Whalley (1987). J.Chem.Res. 30-31, 401-417
C.J. Gilmore, T.R. Rae (1984). Acta Cryst. C 40, 1630-1632
W. Hinriches, G. Klar (1982). J.Chem.Research 336-337, 3540-3593
W. Hinriches, H-J. Riedel, G. Klar (1982). J.Chem.Research 334-335, 3501-35
F. Imashiro, A. Saika, Z. Taira (1987). J.Org.Chem. 52, 5727
O. Johnson, D.W. Jones (1989). Z.Kristallogr. 189, 109-116
O. Johnson, D.W. Jones, J.D. Shaw (1989). In 11th International Symposium on Polynuclear Aromatic Hydrocarbons (eds: K. Loening, M. Cooke, W.E. May). Lewis, Washington, D.C.   In the press
D.W. Jones, C.E. Briant, J.D. Shaw (1988). In Molecular Structure: Chemical Reactivity and Biological Activity (eds: J.J. Stezowski, J-L. Huang, M-C.Shao). University Press, Oxford, pp.221-229
D.W. Jones and J.D. Shaw (1987). Carcinogenesis 8, 1323-1326
D.W. Jones and J.D. Shaw (1989a). Carcinogenesis 10, 1829-1831
D.W. Jones and J.D. Shaw (1989b) Unpublished measurements
C.L. Klein, E.D. Stevens, D.E. Zacharias, J.P. Glusker (1987). Carcinogenesis 8, 5-18
A. Matsuo, H. Nozaki, M. Suzuki, M. Nakayama (1985). J.Chem.Res. 174-175, 1913-1981
G.R. Pettit, S.B. Singh, M.L. Noven, J.M. Schmidt (1988. Can.J.Chem. 66, 406-413
H.W. Schmalle, O.H. Jarchow, B.M. Hausen, K-H. Schulz, K. Krohn, U. Loock (1986). Acta Cryst. C 42, 1039-1040
F.R. Stermitz, T.R. Suess, C.K. Schauer, O.P. Anderson, R.A. Bye, Jnr. (1983). J.Nat.Prod. 46, 417
T.R. Welberry, R.D.G. Jones, M. Puza (1983). Acta Cryst. C 39, 1123-1127

10

The relation between S→O hypervalent bonds and S . . . O close contacts in crystal structures of sulfuranes, analogous sulfonium salts, and related compounds containing X—S . . . O (X=S, Cl, O, N, C) moieties

A. Kálmán

Following the X-ray analysis of a dichlorosulfurane (Baenziger et al, 1969) and the fundamental work of Musher* (1969) which predicted the existence of the first stable diaryldiacyloxyspiro-sulfurane (I), synthesized and structurally described two years later by Kapovits and Kálmán (1971), considerable work has been done on the structures (and chemistry) of various kinds of tetracoordinate tetravalent sulfur species. In particular, Martin and co-workers (Paul et al, 1971; Perozzi et al, 1974; Adzima et al, 1977; Adzima et al, 1978; Lam et al, 1981) published a series of papers on various sulfurane properties, e.g. the polarizability of hypervalent bonds, apicophilicity orders of substituents, etc.

For SF_4 Musher gives a different (perhaps better) interpretation of the long axial bonds than Gillespie's model. In his model, which can also be extended to the sulfuranes, Musher uses only the orthogonal p orbitals of the sulfur atom. He takes the pragmatic position that there is no need to introduce the d orbitals; two ordinary (covalent) bonds are formed in which two singly occupied p orbitals (say p_x and p_y) from the nucleus are bonded to two singly occupied atomic orbitals from two fluorine ligands, respectively. The hypervalent bonds are formed so that the doubly occupied p orbital (say p_z^2) of the central atom is bonded with two singly occupied atomic orbitals from two collinear fluorine ligands. The molecular orbital description of this three-centre-four-electron process involves a doubly occupied bonding and a non-bonding orbital which are written equally well in terms of localized equivalent

---

* 1969-1989: In memoriam Professor Jeremy I. Musher (1935-1974)

orbitals. Musher concluded that, with sulfur as hetero atom, not only chlorine and fluorine are sufficiently electronegative to form hypervalent bonds, but it should also be possible to form such molecules with ligands OR, C_6H_5, CF_3, etc. The stability of the molecules with these "weak hypervalent bonds" can be increased if they are cyclized. This prediction by Musher is supported by the relatively high melting point of the cyclic sulfuranes (mp. of I is 303 $^\circ$C).

The hypervalent S→O bond lengths (1.71-1.96 Å) observed in the first sulfuranes (Kálmán et al, 1973, Adzima et al, 1977 and references therein) not only extended the range of the covalent S-O bond lengths (1.4-1.7 Å) but also developed a bridge to the S(II)...O close contacts as reported first by Mammi and co-workers (1961) and revealed subsequently in numerous crystal structures (Stanković et al, 1980; Kucsman et al, 1984; Kucsman et al, 1985; Kucsman et al, 1989; Párkányi et al, 1989). It has been shown (Kálmán and Párkányi, 1980) that, for these X-S...O=Y no-bond/single--bond interactions (II), depending on the electronegativity of X, the polarity of S-X and the character of the Y moiety, the S...O distances vary from 2.03 Å (Johnson et al, 1971) up to the sum of the van der Waals radii of O and S: 3.25 Å. A report (Lam et al, 1981) on an unsymmetrical sulfurane (III) indicated an overlap between the range of the sulfurane S→O bond lengths and that of the S(II)...O close contacts embedded in a three-centre-two-electron system. However, the difference in the valence of sulfur in models II and III raised doubts about their comparability. The missing link has recently been found by Kapovits and co-workers (Budapest) synthesizing a series of sulfurane-like sulfonium salts represented by IV, V and VI. Their molecular structures determined by X-ray diffraction (Szabó et al, 1990) can be described, like those of sulfuranes, as a distorted trigonal bipyramidal arrangement of the ligands about a sulfonium centre (disregarding the anions). In axial directions the nearly linear O-S$^+$...O= or N-S$^+$...O= array (179+1°) suggests a three-centre-four-electron bond formed by a very strong S$^+$...O= (IV: 2.260(2), V: 2.282(3), VI: 2.373(3) Å, respectively) through-space interaction (sulfur-anium salt character). The anions, being practically in the equatorial plane, approach the sulfonium centre asymmetrically.

X-ray analysis of more than a dozen tetracoordinate tetravalent sulfur compounds (so far unpublished) with asymmetrically and symmetrically enlarged hetero rings incorporating either N-S$^+$...O= or N-S-O bridges revealed that the S...O distances distributed evenly in the range 2.04-2.78 Å are balanced by covalent S-N bonds varying in a narrow range from 1.78 to 1.66 Å. In accordance with the bond order conservation principle, one hundredth Å change in the strong S-N multiple bonds corresponds to an order of magnitude greater alteration in the S...O close contacts. These together suggest a reconsideration of the interpretation of the S→O distances in strongly asymmetrical sulfuranes or analogous sulfonium salts. It seems that if one bond in these three-centre-four-electron systems gains a strength which brings it close to a covalent single bond or even further towards multiple bonds (cf. S-N distances around 1.66-1.68 Å) then the hypervalent character of the other bond rapidly fades out. From this it follows that hypervalent bonds classified by Musher exist almost exclusively in sulfuranes possessing either symmetrical O-S-O (VII, VIII and IX) or N-S-N bridges (Adzima et al, 1978).

Molecular diagrams. I-VI: as referred in the text; VII and VIII: unpublished results of the authors and co-workers (compound VIII sitting on a twofold axis in the space group Fdd2 is quasi isostructural with I); IX: Adzima et al, 1977.

On the other hand, the great variety of S(IV)...O distances observed in these asymmetric sulfuranes and analogous sulfonium salts (the structure interpretations of which are in progress) provides a better understanding of the almost unbroken continuity of the S...O no-bond/single-bond resonance interactions decreasing from 3.25 Å down to the range of the real hypervalent bonds. Thus the common feature of the three--centre-four-electron and the three-centre-two-electron systems can be outlined. These may also help to shed light on the incipient stages of intra- or even intermolecular reactions. Equally it is expected that these structural studies by X-ray diffraction also help elucidate the fiercely debated electronic structures of the hypervalent bonds and the weaker S...O close contacts (Cohen-Addad et al, 1984). Several attempts have been made to improve the classical Musher model of hypervalency. The author is inclined towards Mayer's (1989) conclusion suggesting "that the rather positive charge on the sulfur involved in hypervalent-bond formation is expected to increase the probability that the sulfur d orbitals will also be populated to a considerable extent. The population of d orbitals may, in turn, also have some influence on the hypervalent bonds."

The author wishes to thank Professor I. Kapovits, Drs J. Rábay, D. Szabó (organic chemists) M. Czugler, V. Fülöp, T. Koritsánszky (crystallographers) and I. Mayer (quantum chemist) for their invaluable help in this work. Thanks are also due to Mr. Cs. Kertész and Mrs. Gy. Tóth-Csákvári for their technical assistance.

REFERENCES

ADZIMA,L.J., DUESLER,E.N. and MARTIN,J.C. (1977). J.Org.Chem. 42, 4001.
ADZIMA,L.J., CHIANG,C.C., PAUL,I.C. and MARTIN,J.C. (1978). J.Amer.Chem.Soc. 100, 953.
BAENZIGER,N.C., BUCKLES,R.E., MANER,R.J. and SIMPSON, T.D. (1969). J.Amer.Chem.Soc. 91, 5749.
COHEN-ADDAD,C., LEHMANN,M.S., BECKER,P., PÁRKÁNYI, L. and KÁLMÁN,A. (1984). J.Chem.Soc.Perkin.Trans. 2, 191.
JOHNSON,P.L., REID,K.I.G. and PAUL,I.C. (1971). J.Chem.Soc. B, 946.
KÁLMÁN,A., SASVÁRI,K. and KAPOVITS,I. (1973). Acta Cryst. B29, 355.
KÁLMÁN,A. and PÁRKÁNYI,L. (1980). Acta Cryst. B36, 2372. (and references therein)
KAPOVITS,I. and KÁLMÁN,A. (1971). J.Chem.Soc.Chem. Commun. 649.

KUCSMAN,Á., KAPOVITS,I., PÁRKÁNYI,L., ARGAY,GY.
and KÁLMÁN,A. (1984). J.Mol.Struct. 125, 331.
KUCSMAN,Á., KAPOVITS,I., KÖVESDI,I., KÁLMÁN,A. and
PÁRKÁNYI,L. (1985). J.Mol.Struct. 127, 135.
KUCSMAN,Á., KAPOVITS,I., CZUGLER,M., PÁRKÁNYI,L.
and KÁLMÁN,A. (1989). J.Mol.Struct. 198, 339.
LAM,W.Y., DUESLER,E.N. and MARTIN,J.C. (1981).
J.Amer.Chem.Soc. 103, 127.
MAMMI,M., BARDI,R., TRAVERSO,G. and BEZZI,S. (1961).
Nature 192, 1282.
MAYER,I. (1989). J.Mol.Struct.(Theochem.) 186, 43.
MUSHER,J.I. (1969). Angew.Chem.Int.Ed.Engl. 81, 68.
PÁRKÁNYI,L., KÁLMÁN,A., KUCSMAN,Á. and KAPOVITS,I.
(1989). J.Mol.Struct. 198, 355.
PAUL,I.C., MARTIN,J.C. and PEROZZI,E.F. (1971).
J.Amer.Chem.Soc. 93, 6674.
PEROZZI,E.F., MARTIN,J.C. and PAUL,I.C. (1974).
J.Amer.Chem.Soc. 96, 6735.
STANKOVIC,S., RIBÁR,B., KÁLMÁN,A. and ARGAY,GY.
(1980). Acta Cryst. B36, 1235.
SZABÓ,D., KAPOVITS,I., KUCSMAN,Á., FÜLÖP,V.,
CZUGLER,M. and KÁLMÁN,A. (1990). Struct.Chem.
1, 305.

11

ISIS: a powerful new tool for organic crystal chemistry

J. L. Finney and C. C. Wilson

1 INTRODUCTION

Crystallographic methods to solve crystal and molecular structures are central to organic crystal chemistry. X–rays have generally been the preferred probe for crystallographic work due to the wide availability and convenience of laboratory X–ray sources. Most of the structural work to date has been performed on single crystals, which have been capable of giving data of much higher quality than powder (polycrystalline) materials. Also important in this field are spectroscopic methods, which can be used to improve our understanding of the interactions between molecules. In spectroscopy, infrared and Raman, along with NMR, have been extensively exploited.

Alongside these techniques, the use of neutrons, both for crystallographic structural work and inelastic spectroscopy, has also developed. Because beams of neutrons are produced only at large central facilities, their use has been less widespread than the more conventional techniques which can be carried out in one's own laboratory. However, the development over the last two decades of advanced high flux neutron sources and instrumentation such as that at the Institut Laue–Langevin (ILL), Grenoble, France, has led to a growing user community exploiting the particular advantages of neutrons in structural and dynamical studies of molecules in crystals.

Many of the advantages of neutrons over X–rays relate to the nature of the neutron–nucleus interaction quantitatively described by the neutron scattering factor, or scattering length b. Briefly, these can be summarised as follows :

a) Unlike X–rays, which are scattered by **electrons**, neutrons are scattered by the point nucleus. This results in the neutron scattering length being independent of scattering vector S ($\sin\theta/\lambda$) or Q ($4\pi\sin\theta/\lambda$). Consequently, unlike with X–rays, neutron data can be taken to very high Q, leading to very high real space resolution.

b) For X–rays, the atomic scattering factor relates directly to the number of electrons. In contrast, the neutron scattering length is an irregular function of atomic number. Thus, light atoms may scatter as strongly – or more strongly – than heavy ones. Neutrons are thus particularly useful where light atoms are of interest – especially in the presence of heavy ones, a situation which is awkward for X–rays. Neutrons in particular are good for looking at **hydrogen** – an important atom in organic molecules!

c) As the neutron scattering factor depends on the **nucleus**, different isotopes of the same element may scatter differently. This opens up the possibility of **contrast variation** or **difference** measurements using isotope substitution – a technique which can be very powerful in particular applications.

d) The neutron **spin** can interact with the spin of **electrons**, and hence can be used to probe magnetic structures.

In addition to having these scattering factor characteristics, neutrons tend to be only weakly absorbed by samples. This allows us to look at samples in quite massive sample environment chambers (e.g. pressure cells, furnaces, cryostats, reaction vessels), and also means that molecules containing heavy atoms are no problem. Finally, the typical energy of neutrons of wavelengths suitable for diffraction measurements ($\sim$ Å) is of the same order as those of thermal excitations in molecules. Thus, by measuring the change in energy on inelastically scattering a neutron, we can probe the **dynamics** of molecules using neutrons from the same source as those used to probe structure (occasionally structure and dynamics are measured simultaneously!)

2 RECENT DEVELOPMENTS IN NEUTRON PRODUCTION: FROM REACTORS TO PULSED SOURCES

The past 20 years or so have seen increased exploitation of these characteristics of neutron scattering, most of it on reactor sources, where neutrons are produced by fission of uranium. As demands from experimentalists have increased, pressure has risen to develop further neutron sources to higher fluxes. However, we appear to be approaching the limit of development of reactor sources, limits which seem to be set by engineering considerations related to the problems of power density in the reactor core. These technological limitations are indicated by the fact that the "next generation" of high flux reactors is postulated to have fluxes of some 5×10^{15} n.cm^{-2}.s^{-1} compared with the present $> 10^{15}$ flux of ILL, an increase by a factor of only around four. Other methods of neutron production which are not so limited have therefore been sought.

One particularly promising way forward is to produce neutrons by the process of **spallation** of heavy atom nuclei by bombarding a heavy atom target with pulses of protons from a high energy accelerator. The world's most powerful such pulsed spallation neutron source is ISIS, situated at the Rutherford Appleton Laboratory of the UK Science and Engineering Research Council. From its first neutron production in December 1984, it has now built up an extensive experimental programme over a wide range of condensed matter science.

In ISIS (Figure 1), pulses of protons are injected at 70 MeV into a synchrotron ring. These are then further accelerated (presently to 750 MeV), and then kicked out into an extracted proton beam line, at the end of which is a heavy metal target (presently depleted U or Ta). On hitting the target, each proton produces (for uranium) around 25–30 neutrons. As in a reactor, however, these neutrons are of energies too high to be of use in condensed matter studies. They are therefore allowed to interact with one of four moderators held at different temperatures which reduce their energies before being conducted along a beam tube to one of the array of instruments sited around the target station (Figure 2).

In a reactor source, the neutron production is continuous, and the moderation process is allowed to go to equilibrium. Thus, the neutron spectrum issuing from a moderator follows a Maxwellian distribution that is characteristic of the moderator temperature. In most **constant wavelength** (CW) measurements made on reactors, a narrow monochromatic beam is then selected from the Maxwellian distribution by using either a crystal monochromator or rotating chopper.

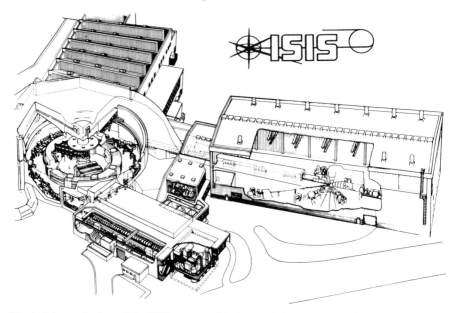

Fig. 1 Schematic view of the ISIS source, with the pre–injector and the linac at the bottom, proton synchrotron on the left, and the target station and instruments in the experimental hall on the right.

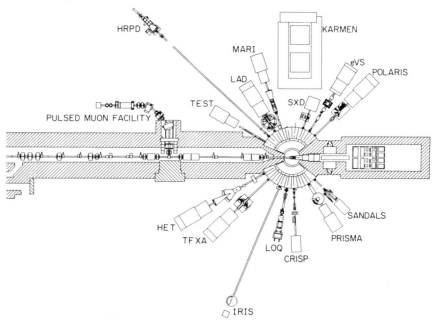

Fig. 2 Layout of the ISIS instruments.

On a pulsed source, the procedures are significantly different. Central to their efficient operation is the **time structure** of the neutron pulse (see below); were we to allow complete moderation, the time structure would be lost. The moderation is therefore incomplete and the resulting spectrum has a high energy (short wavelength) component of **epithermal** neutrons, in addition to the Maxwellian.

This epithermal neutron component is one of the major advantages of pulsed spallation sources over reactors: the presence of a high flux of short wavelength neutrons allows us to perform elastic measurements to very high values of $\sin\theta/\lambda$, and hence to obtain very high resolution in real space. Additionally, high energy inelastic excitations can be probed.

The other major advantages of pulsed spallation neutrons relate directly to the pulsed nature of the neutron beam. These can be summarised as follows :

a) Because the neutrons come in pulses (at ISIS at 50 Hz) off the moderator, we know within a small error the time (t_O) at which all our neutrons are "created" in the moderator. As we also know the time at which our neutron is finally detected (t_D), we can use this **time–of–flight** (t_D-t_O) to measure the energy of our scattered neutron. This is simply because the speed of the neutron depends on its energy, and hence its wavelength: knowing both the time–of–flight and the flight path, we therefore know the neutron wavelength.

b) The neutron spectrum emerging from the moderator is a white spectrum covering a range of wavelengths. Using time–of–flight, we can therefore use in principle every neutron in an experiment since each arrives at the detector with a tag (time) representing its wavelength. Without the neutrons being pulsed we would in general have to monochromate as in a reactor. With the time structure, we can make full use of the white beam. Pulsed sources thus make much more efficient use of neutrons.

c) A powerful consequence of using time–of–flight is that the **resolution** of (elastic or inelastic) experiments increases with flight path, and can therefore be made in principle arbitrarily high. Moreover, in a crystallographic experiment, the resolution, defined as $\Delta d/d$, is essentially constant with d. This is **not** true for a reactor source, in which, using a crystal monochromator, resolution deteriorates at both low angles relevant to e.g. magnetic structures, and high angles where good data is needed for high quality refinements.

d) Using a white incident beam favours experiments for which a wide dynamic range (e.g,. in Q or energy transfer) is advantageous. In fact, we can in principle obtain a complete diffraction pattern from a polycrystalline material at a single scattering angle. Fixed geometry work is therefore possible, and can be used to great advantage at 90°, an angle which allows scattering from sample environment such as pressure cells to be collimated out.

e) Pulsed sources have inherently very low backgrounds: the source is essentially switched off when the data is being collected.

The development potential for pulsed sources is at present significantly greater than for reactors. The ISIS current is at present some 110μA, and it is intended to approach 200μA in the next three years. Designs for the "next generation" of pulsed sources reach some 4000μA at 1600MeV, a potential increase in neutron flux of >80 times over the current highest (cf. factor of no more than 4 for reactors).

Although it is still early days for pulsed neutrons (the ISIS scientific programme is essentially little more than three years old), users of such sources have begun to capitalise on these characteristics. In the remainder of this paper, some examples of the use of ISIS in selected areas relevant to organic crystal chemistry will be briefly described.

3 HIGH RESOLUTION POWDER DIFFRACTION

The ISIS high resolution powder diffractometer HRPD (Figure 3; Johnson and David, 1985) is the highest resolution neutron powder diffractometer in the world. Sitting at the end of a beam line almost 100m in length,, and detecting in backscattering geometry, a resolution of $\Delta d/d \sim 5 \times 10^{-4}$ is achieved. Moreover, this resolution is essentially constant over the wide d–spacing range available (currently down to d–spacings of less than 0.3Å) making a much larger number of reflections available – and at higher resolution – than is possible on any other neutron powder instruments. Figure 4 shows data from a powder of p–xylene

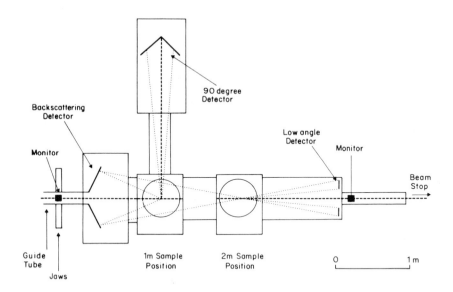

Fig. 3 Plan view of HRPD. The 2m sample position is some 96m from the moderator.

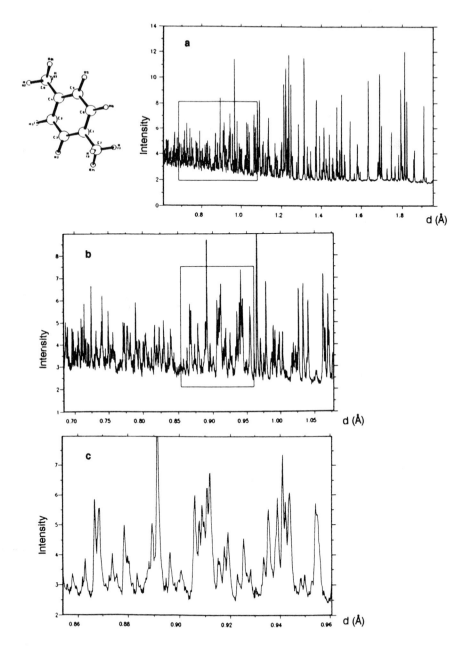

Fig. 4 Data collected on HRPD from p–xylene (P2$_1$/n, a = 5.7344, b = 4.9495, c = 11.1389Å, β = 100.71°) in the d–spacing ranges 0.6–2.0Å (a), 0.68–1.08Å (b) and 0.85–0.96Å (c).

(Prager, 1989); the quality of the data is emphasised by the zoom windows. HRPD also exploits the high source brightness and low background, improved further by the use of a curved guide to remove fast neutrons from the instrument, and has a detector bank at 90° for confined sample environment work.

HRPD has dramatically extended the range of problems that can be tackled by powder diffraction. For example :

- Rietveld profile refinement of more complex structures are possible;
- Hydrogenous samples – often a problem for neutrons because of the high incoherent background – can be tackled;
- Crystal structures can be solved from powders *ab initio*;
- Structure refinements can be performed of a quality which previously required single crystals;
- The high resolution allows sophisticated peak shape analyses enabling the study of, for example, texture and strain.

Examples of *ab initio* structure determination include the high temperature α–phase of malonic acid (Figure 5; Delaplane,1989). The α-phase is reached by a first order phase transition, which excludes the possibility of single crystal techniques. From data collected at 100°C on a deuterated sample over a d-spacing range of 0.6–5.2Å, 290 reflections were extracted. Both cell determination by autoindexing and structure solution by direct methods followed routinely. Table 1 gives the result of the structure determination and illustrates the quality of the refinement. The routine nature of this *ab initio* structure determination is indicative of the power of HRPD in developing and exploiting the new opportunities for performing crystallography that have been presented to us by pulsed neutron sources.

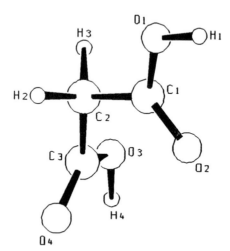

Fig. 5 The malonic acid molecule, the high temperature phase of which was solved *ab initio* from HRPD data.

Table 1 Malonic acid structure determination and refinement information

Malonic acid, α-$CD_2(COOD)_2$

Structure determination :

Auto-indexing using 47 reflections in TREOR (Werner et al, 1985),

Cell determined : Orthorhombic, a = 7.641, b = 5.015, c = 11.539Å

Space group from absences : Pbcn

Structural solution :

287 reflections extracted, 55 unobserved, 232 used in solution attempt by direct methods

83 reflections with $|E_h|$ > 1 used by MITHRIL (Gilmore, 1984) in routine solution using just TRIPLETS. Most atoms indicated in first E-map :

Peak number	X	Y	Z	Height	atom
1	0.5174	0.2031	0.6436	1721	C(2)
2	1.0000	0.1388	0.2500	1527	C(1)
3	0.1285	0.4687	0.1313	1180	O(2)
4	0.6413	0.2739	0.5763	1018	O(1)
5	0.1352	-0.0071	0.2504	922	D(2)
6	0.3688	0.1941	0.1623	773	
7	0.8834	0.4159	0.5224	701	
8	0.1730	0.2879	0.5246	643	

Structural refinement :

Cell parameters : a = 7.64027(1), b = 5.01597(1), c = 11.54623(1)Å

Atomic parameters :

Atom	X	Y	Z	B_{iso} (Å^2)
C(1)	0.5000	0.3901(9)	0.7500	4.7(1)
C(2)	0.5072(5)	0.2027(6)	0.6451(3)	3.9(1)
O(1)	0.6173(7)	0.2843(12)	0.5665(7)	7.7(2)
O(2)	0.4005(5)	0.01856(8)	0.6297(3)	4.8(1)
D(1)	0.3941(8)	-0.1644(14)	0.5045(7)	*
D(2)	0.3724(6)	0.4801(7)	0.7473(3)	6.9(1)

* Atom D(1) possibly disordered, thermal parameters unstable

J. L. Finney and C. C. Wilson

An example of powder work on hydrogenous systems is the very accurate determination of the structure of squaric acid $C_4H_2O_4$ (Nelmes, Tun, David and Harrison, 1987). The original experiments at room temperature using single crystal neutron and X-ray scattering showed that the molecule possesses a planar structure (Figure 6a) which has a very small monoclinic distortion from tetragonal symmetry. The accurate location of hydrogen in a material such

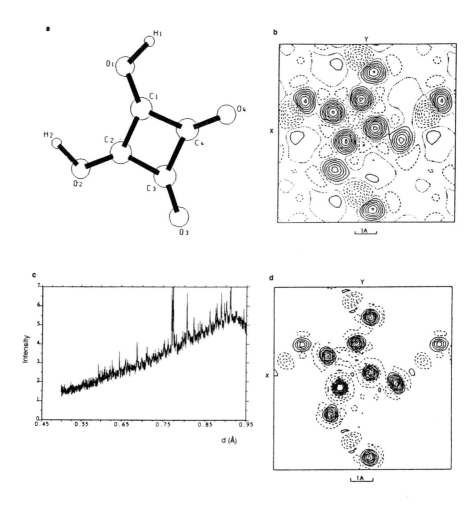

Fig. 6 (a) The squaric acid molecule; (b) Fourier map calculated from HRPD data to 1Å resolution; (c) The raw HRPD data below 0.95Å; (d) Fourier map calculated at 0.54Å data resolution, showing the improvement gained.

as this is a good test of the capabilities of improved resolution to allow these parameters to be obtained. The data were recorded in 12 hours from a sample of some 1.5g and profile refinement gave an excellent fit. As can be seen from the Fourier map calculated using these data ($d_{min} = 1$ Å) the hydrogen atoms are clearly indicated (Figure 6b). The hydrogen parameters found in the powder study agree well with the neutron single crystal study and are substantially better determined than in the single crystal X-ray study. The lattice parameters obtained in the refinement indicate the very high accuracy obtained on HRPD ($a = 6.12890(5)$, $b = 5.26781(5)$, $c = 6.14025(6)$ Å, $\beta = 89.9632(5)°$).

Subsequent to the refinement of this data set, a further set was collected which extended the d–spacing range down to d_{min} of 0.54Å (Figure 6c). The dramatic improvement in the Fourier map calculated including this latter data (Figure 6d) shows the value of improved resolution in the diffraction pattern and shows the capability of HRPD to operate successfully at very high Q values. The well–defined atoms in the high resolution Fourier map allowed an analysis of the order–disorder character of the O–H...O hydrogen bond between neighbouring squaric acid molecules which, as expected, was found to be fully ordered at room temperature.

A final example of the power of HRPD is given by recent work on benzene, C_6H_6 (David, 1989). Benzene adopts an orthorhombic structure, space group Pbca ($Z = 2$), with a moderately-sized unit cell ($a = 7.3550$, $b = 9.3709$, $c = 6.6992$Å, $V = 461.7$Å^3) (Figure 7a). Although the original structure determination (Cox, 1928) located only the carbon atoms, the observed molecular planarity resolved a debate about whether the molecule was puckered, as favoured by a number of eminent scientists including Bragg, or flat.

Successive X–ray single crystal investigations improved the precision and accuracy of structure determination. The use of neutrons as a structural probe confirmed the planarity not only of the carbon but also the hydrogen atoms. A recent single crystal neutron diffraction investigation on deuterated benzene, C_6D_6 (Jeffrey, Ruble, McMullan and Pople, 1987), sought to investigate any deviations from planarity. In addition to contributing to continued investigations of the nature of the chemical bond, these experimental results have been compared with the latest theoretical calculations to assess the current status both of experimental technique and theoretical calculation. One further justification was given by Jeffrey for performing another single crystal neutron diffraction experiment: benzene "should be repeatedly investigated by each scientific method whenever there is a significant advance with respect to the detail or accuracy that the method can offer". It is therefore fitting that a detailed structural investigation of benzene be performed on HRPD to investigate (i) how well do the best powder diffraction experiments compare with equivalent single crystal studies for moderately complex structures with unit cells of the order of 500Å^3 and (ii) can reliable anisotropic temperature factors be obtained from time–of–flight powder diffraction experiments? The existence of both high precision single crystal data and detailed theoretical calculations provides a very rigorous test for the benzene data collected on HRPD.

Data were collected over a period of approximately nine hours (174 μA–hr). The raw data were corrected for incident flux (using a vanadium calibration), and cryostat and sample attenuation. The last correction was derived from consideration of the transmitted neutron flux and showed significant structure from multiple-scattering self-attenuation effects. Rietveld refinement was performed using the powder diffraction package developed at RAL and based upon the Cambridge Crystallography Subroutine Library (Brown and Matthewman, 1987). For data in the range 0.606–1.778Å an excellent least-squares fit to the powder diffraction data was obtained (Figure 7b). The refined structural parameters, including 18 atomic coordinates and 36 anisotropic temperature factors, are listed in Table 2. Experimental and theoretically–calculated anisotropic temperature factors are presented in Table 4. With the exception of the B_{22} temperature factors for three carbon atoms, there is a remarkable agreement between the temperature factors obtained from HRPD and from the single crystal data. More importantly, the anisotropic temperature factors for the deuterium atoms calculated using harmonic lattice dynamical calculations (Filippini and Gramaccioli 1989) are significantly different from those obtained by powder and single

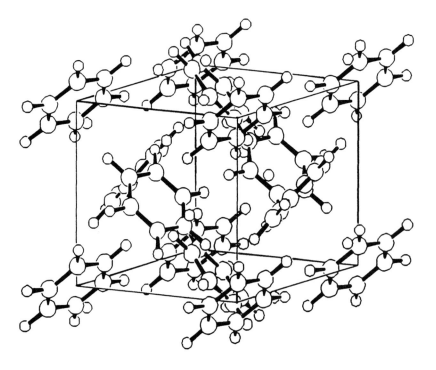

Fig. 7 (a) View of the crystal structure of benzene

crystal diffraction techniques. Table 3 lists bond lengths, uncorrected for libration, obtained in the present study and from the work of Jeffrey et al (1987). The agreement is good, with few statistically significant differences, these probably resulting from systematic errors in the powder diffraction data.

Table 2 Refined structural parameters for benzene from HRPD

space group Pbca, $Z = 2$, molecular symmetry $= \bar{1}$

$a = 7.3551(3)$ Å, $b = 9.3712(4)$ Å, $c = 6.6994(3)$ Å, $V = 461.76$ Å^3

	x/a	y/b	z/c	B_{iso} (x 10^{-4} Å^2)
C1	−0.06120(15)	0.14123(10)	−0.00519(20)	68(6)
C2	−0.14023(15)	0.04469(10)	0.12722(15)	66(6)
C3	−0.07770(15)	−0.09689(12)	0.13264(20)	77(6)
D1	−0.10853(15)	0.25050(15)	−0.01187(25)	202(9)
D2	−0.24908(20)	0.07682(15)	0.22600(20)	202(9)
D3	−0.13821(20)	−0.17136(15)	0.23703(20)	203(9)

$R_P = 13.2\%$ $R_{wP} = 15.7\%$ $R_E = 10.8\%$ $\chi^2 = 2.1$

Table 3 Benzene bond lengths (uncorrected for libration)

HRPD data (4 K) (present work)

C1–C2 = 1.3940(20) Å	C1–D1 = 1.0825(30) Å
C2–C3 = 1.4047(30) Å	C2–D2 = 1.0815(30) Å
C1–C3 = 1.3948(20) Å	C3–D3 = 1.0836(25) Å
mean = 1.3978(15) Å	mean = 1.0825(20) Å

Single crystal neutron diffraction (15 K) (after Jeffrey et al 1987)

C1–C2 = 1.3969(7) Å	C1–D1 = 1.0879(9) Å
C2–C3 = 1.3970(8) Å	C2–D2 = 1.0869(9) Å
C1–C3 = 1.3976(7) Å	C3–D3 = 1.0843(8) Å
mean = 1.3972(5) Å	mean = 1.0864(7) Å

J. L. Finney and C. C. Wilson

Table 4 Benzene at low temperatures: anisotropic temperature factors (x 10^4 $Å^2$)

The four lines represent : Harmonic lattice dynamical calculation at 15K (1) and 0K (2) (Filippini and Gramaccioli, 1989); Neutron single crystal data at 15K (3) (Jeffrey et al, 1987); Neutron powder diffraction data at 4.2K (David et al, 1989)

Atom	B_{11}	B_{22}	B_{33}	B_{23}	B_{13}	B_{12}
C1	90	66	89	3	7	0
	77	58	77		0	7
	79(2)	67(2)	88(2)	4(1)	7(2)	6(2)
	77(7)	42(6)	87(7)	1(5)	5(5)	–3(4)
C2	84	87	82	–2	17	6
	71	79	70	–2	17	9
	74(2)	81(2)	79(2)	0(2)	17(2)	9(2)
	71(7)	58(7)	68(6)	9(4)	26(5)	12(4)
C3	86	79	82	10	11	6
	73	72	70	10	11	–5
	81(2)	75(2)	82(2)	10(1)	14(2)	–3(2)
	83(7)	57(7)	92(7)	0(5)	18(5)	–1(4)
D1	216	173	212	11	39	38
	222	165	199	11	39	38
	224(3)	114(2)	239(3)	12(2)	25(2)	46(2)
	218(8)	121(7)	267(9)	19(5)	22(6)	31(5)
D2	184	226	215	5	56	56
	170	217	202	5	55	56
	183(2)	204(3)	208(3)	–8(2)	88(2)	35(2)
	170(8)	212(8)	225(9)	–2(6)	120(6)	33(5)
D3	228	208	171	60	37	–2
	215	199	158	59	36	–1
	214(3)	171(2)	199(3)	58(2)	61(2)	–18(4)
	241(9)	155(8)	214(8)	75(6)	66(7)	–20(5)

At the present stage of analysis it is clear that the end results obtained from refinement of HRPD data are not significantly inferior to the **best** single crystal data. Both experimental techniques agree closely with each other, and differ from the theoretical calculations, particularly in the values obtained for the anisotropic temperature factors for the deuterium atoms. The powder diffraction experiment thus strongly supports the single crystal study and indicates that further improved theoretical calculations are required. The quality of these powder diffraction results represents present state of the art at ISIS. Further improvements in the normalisation procedure are currently under development and should lead to a precision and accuracy in moderately complex structure determination that is at least as good as the best single crystal results. This indeed represents a very significant advance in the power of powder diffraction.

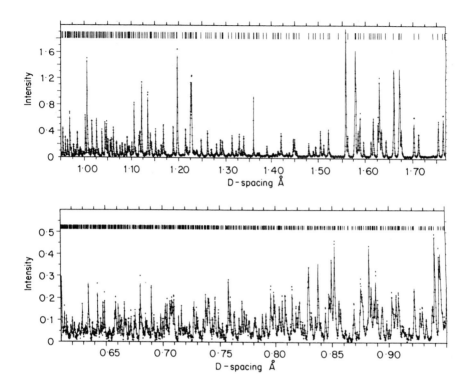

Fig. 7 (b) The final profile fit of the HRPD low temperature benzene data, from which accurate anisotropic temperature factors were obtained. The tags at the top represent the positions of Bragg reflections. The quality of the fit is apparent.

4 HIGH INTENSITY POWDER DIFFRACTION

The recently commissioned medium resolution powder diffractometer POLARIS is an instrument that is in many ways complementary to HRPD. Being situated considerably closer to the ISIS target, it sacrifices some resolution in favour of high count rates. The ability thus engendered to study smaller or more weakly scattering samples can be very relevant to organic chemistry applications. The loss of resolution is important, of course, especially when a material with a large, low symmetry unit cell, and hence many Bragg peaks in the pattern, is being studied. To some extent, however, this problem is offset by the availability at ISIS of extremely powerful and flexible profile refinement packages.

In addition to having the capability to perform measurements on small samples over normal times, POLARIS also gives the possibility of taking data over very short times on samples of standard size. It is thus an appropriate instrument on which to look in real time at phase changes and solid state chemical reactions. Furthermore, the existence of a large detector bank at 90° allows high quality measurements on samples in pressure cells or other sample environment. A recent example is a data set taken on ice VI at 5.5 kbar and 230 K (Finney, Kuhs and Londono, unpublished work). Figure 8 shows the diffraction pattern at 90°, compared with simulated data: no peaks from the sample environment can be seen. Before the scattering from the pressure cell had been collimated out, they dominated the measured pattern, some peaks being orders of magnitude higher than those from the sample. This illustrates the power of using fixed sample geometry on pulsed sources; in a conventional constant wavelength scan on a reactor source, eliminating the lines from the pressure cell can be a major problem.

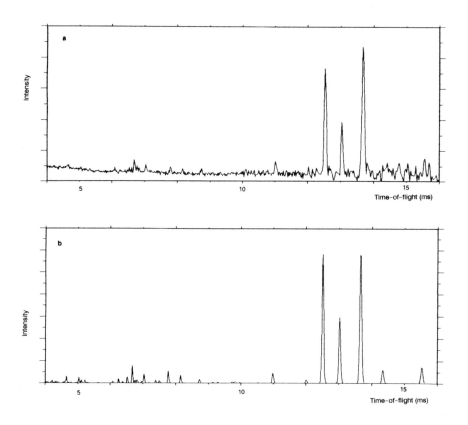

Fig. 8 Ice VI data collected at high pressure and low temperature on POLARIS (a), compared with simulated data (b). The absence of spurious peaks from the pressure cell is clear.

5 SINGLE CRYSTAL DIFFRACTION

The ISIS single crystal diffractometer SXD uses the time–sorted Laue technique. The instrument is equipped with a position–sensitive detector which, with the addition of the third dimension of time–of–flight on a pulsed neutron source, provides a means of examining large volumes of reciprocal space in one measurement. This surveying ability can be absolutely vital in applications where one is looking for features which occur between Bragg reflections. With an instrument such as SXD there is no need to perform elaborate scans nor indeed (and this is perhaps most important) to know for what one is looking. This large volume surveying approach makes SXD an ideal tool for examining effects such as thermal (TDS) and other diffuse scattering (see Figure 9) which may have a significant effect in structure refinement but which remains largely undetermined in a typical step–scan approach. The improvements in handling TDS which should come from the close examination of this on an instrument such as SXD should help to improve the precision of refinements in a whole range of materials. Thus a measurement rooted solidly in the fundamental physics of the scattering process and made simpler to measure by the SXD technique should be a means towards better and more precise determination of organic structures.

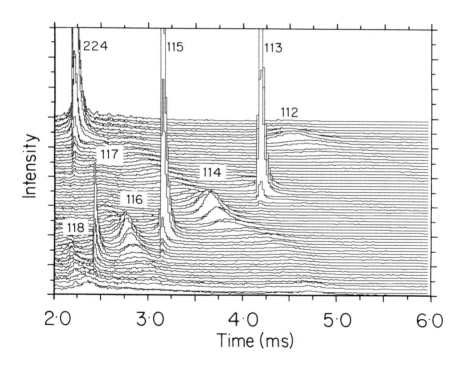

Fig. 9 Bragg and diffuse scattering in a portion of the (hhl) plane of ZrO_2–9.4%Y_2O_3, measured in the equatorial plane of the area detector on SXD. The Laue nature of the SXD instrument makes it ideal for the study of, and subsequent correction for, diffuse scattering effects.

As a routine diffractometer SXD is also a tool for structure determination in its own right. The complementarity of neutrons and X-rays is once again highlighted in one type of application of neutron single crystal diffraction in organic crystal chemistry. Whereas X-rays are scattered from the electron cloud surrounding the nucleus, neutrons interact with the nucleus. The precise determination of the nuclear structure by neutron diffraction can be coupled with an accurate X-ray study to enable the calculation of an X-n difference map, analogous to a normal ($\Delta\rho$) difference Fourier, which should reveal features such as the distortion of the electron density along chemical bonds in a much more relevant way since the nuclear positions are known.

As an example of the high precision of nuclear measurements possible on SXD, anharmonic displacements of the fluorine atoms in the inorganic structure SrF_2 were measured at greater precision than previously obtained (Forsyth, Wilson and Sabine, 1989), due to the very high $\sin\theta/\lambda$ values reached in the experiment. Such a precise understanding of thermal displacements is important in elucidating the overlying electronic structure more precisely.

6 A BRIEF WORD ON NEUTRON SPECTROSCOPY

The dynamics of molecules in crystals are determined by the nature of the interactions between the various atoms in the crystal. Thus spectroscopic information can be used to help understand the interactions between atoms and molecules, basic information in physics, chemistry and biology.

Neutrons are a powerful probe of molecular dynamics through inelastic scattering. The suite of inelastic ISIS instruments allows us to probe, for example :

- incoherent inelastic scattering, especially from hydrogen atoms (HET and TFXA instruments)
- coherent inelastic scattering (phonons) (PRISMA instrument)
- tunnelling and quasielastic studies (IRIS instrument)
- recoil spectroscopy (eVS)

Briefly, recoil spectroscopy allows the measurement of momentum distributions which relate directly to the potential well in which e.g. a hydrogen atom is sitting. Tunnelling spectroscopy of e.g. $-CH_3$ groups gives very sensitive information on the environment of the methyl group, while quasielastic scattering gives information on (translational and rotational) diffusive modes. Coherent inelastic scattering can also be used to test interaction potential functions through comparing measured phonon dispersion curves against those obtained on the basis of assumed potentials.

Of particular interest here is incoherent inelastic neutron spectroscopy. Neutrons are a particularly appropriate tool for investigating interactions between molecules in crystals for several reasons. First, the incoherent neutron scattering cross section of hydrogen is so large that the hydrogen motions dominate the spectrum of hydrogen–containing molecules. Secondly, unlike Raman and infrared, there are no selection rules: **all** modes are in principle observable. Thirdly, the **intensity** of a vibrational line has a simple relationship to the motion

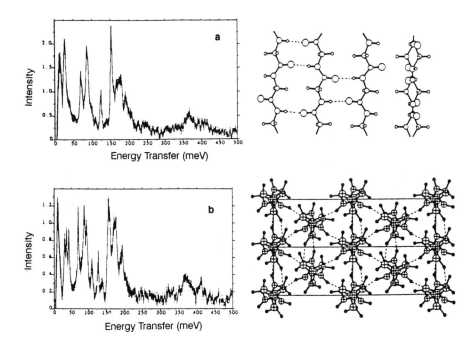

Fig. 10 (a) Spectrum obtained on TFXA from the antiparallel rippled sheet polyglycine I structure (shown); (b) TFXA spectrum from the antiparallel helical polyglycine II (also shown). The differences in the two spectra are very apparent.

of the participating **nuclei**, and can be easily calculated from the amplitude of vibration weighted by the neutron scattering length. This is in contrast to Raman and infrared, where intensities relate to more complex polarisability and dipole moment changes.

Inelastic incoherent spectroscopy on ISIS takes advantage in particular of the low background and intrinsically high resolution available. Figure 10 shows typical spectra taken on the time focused crystal analyser instrument TFXA, and underlines both the high resolution and the wide energy transfer range than can be obtained in a single measurement. The spectra are of two different arrangements of polyglycine: I refers to an antiparallel rippled sheet structure, while II contains antiparallel helices (Baron, Fillaux and Tomkinson, 1989). The observed frequencies are in perfect agreement with Raman and infrared experiments. However, the additional information in the neutron spectra on proton displacements in each mode allows a quantitative analysis of intensities. From this, we conclude quite clearly that the force fields obtained from infrared and Raman need to be significantly modified. A further conclusion is that the $C=N$ torsion mode is strongly affected by the secondary structure (note the strong differences between the spectra of the two crystals between 25–50 meV ($\sim$ 200–400 cm^{-1})). These differences are too large to be explained by anharmonic perturbations.

The potential of such high resolution neutron spectra are only now – with the advent of pulsed spallation sources – beginning to be exploited. One particularly exciting development is that we are now beginning to be able to use spectral profiles to **refine** force fields, by allying the high resolution and the relative ease of calculating intensities from an assumed model. The situation is perhaps similar to that in powder diffraction when profile refinement of crystal **structure** was being developed. With spectra of the quality of Figure 10 (which can be further improved with instrument modifications) we are approaching the stage at which we can profile refine interatomic force constants.

7 SUMMARY

Although, as stated above, the scientific programme on ISIS is little more than three years old, pulsed spallation neutrons have already demonstrated that they can offer much to organic crystal chemists, both in terms of structure determination and refinement, and spectroscopic studies of force fields. The epithermal component of the neutron spectrum allows very high $\sin\theta/\lambda$ data to be collected, while time–of–flight detection facilitates both unrivalled resolution which is essentially constant with d–spacing, and wide dynamic ranges in energy or d–spacing. These advantages now permit single crystal quality data to be obtained from powders, and *ab initio* determination of structure from powder data. Hydrogenous systems – often a problem on reactor sources – can be studied with confidence. High intensity powder diffraction capitalises on source brightness, and enables time–dependent studies or measurements on small samples. The fixed scattering geometry allowed by white beam work has advantages for studying, e.g. samples under pressure. Reciprocal space surveying – e.g. to examine thermal diffuse scattering – is ideally suited to the time–sorted Laue single crystal diffractometer SXD, which also exploits the epithermal spectrum to obtain data to high $\sin\theta/\lambda$ values.

Inelastic neutron scattering can be used to probe the potential functions that control intermolecular interactions. The relatively straightforward relationship of vibrational amplitude and neutron scattering length to the spectral intensity, allied to the high resolution and wide energy transfer range available, promises to make profile refinement of force fields a reality.

The future exploitation of ISIS in this and other fields should continue to increase, with a widely based and expanding scientific user community. Although core funding for ISIS comes from the UK Science and Engineering Research Council, bilateral agreements have been signed with several countries, many of them in Europe. We welcome proposals for the use of ISIS beams from organic crystal chemists and others.

8 REFERENCES

Baron M H, Fillaux F and Tomkinson J (1989) *ISIS 1989 Annual Report*, RAL–**89–050**, p A134.

Brown P J and Matthewman J C (1987). *Rutherford Appleton Laboratory Report* RAL–**87–010**.

Cox E G (1928) *Nature* **122** 401.

David W I F (1989) *ISIS 1989 Annual Report*, RAL–**89–050**, pp 43–46.

Delaplane R G (1989) *ISIS 1989 Annual Report*, RAL–**89–050**, p A5.

Fillipini G and Gramaccioli C M (1989). *Acta Cryst.* A**45** 261–263.

Forsyth J B, Wilson C C and Sabine T M (1989). *Acta Cryst.* A**45** 244–247.

Gilmore C J (1984) *J. Appl. Cryst.* **17** 42–46.

Jeffrey G A, Ruble J R, McMullan R K and Pople J A (1987). *Proc. Roy. Soc. London* A**414** 47–57.

Johnson M W and David W I F (1985). *Rutherford Appleton Laboratory Report* RAL–**85–112**.

Nelmes R J, Tun Z, David W I F and Harrison W T A (1987). In *Chemical Crystallography with pulsed neutrons and synchrotron radiation*, edited by M A Carrondo and G A Jeffrey, p. 479, Proceedings of NATO ASI.

Prager M (1989) *ISIS 1989 Annual Report*, RAL–**89–050**, p A32.

Werner P–E, Eriksson, L and Westdahl, M (1985) *J. Appl. Cryst.* **18** 367–370.

12

Aryloxide-aluminium π-bonding in the π-face selectivity of co-ordinated ketones towards nucleophilic attack

Andrew R. Barron

One of the most important C-C bond formation reactions in organic synthesis is nucleophilic addition to organo-carbonyl groups. A significant development in this area has been the use of the bulky organoaluminum compound, AlMe(BHT)$_2$ (1), for controlling the π-face selectivity of carbonyl moieties towards nucleophilic addition (Maruoka, Itoh, Yamamoto, 1985; Maruoka, Itoh, Sakuria, Yamamoto, 1988).

(1)

In this system, the stereo-selective equatorial (anti-Cram) alkylation of substituted cyclohexanones was believed to occur by complexation of AlMe(BHT)$_2$ preferentially to one π-face of the ketone (I).

I

On the basis of <u>ab initio</u> calculations (LePage & Wiberg, 1988) a more consistent geometry of the ketone-aluminum complex has aluminum in the π-nodal plane of the ketone **(II)**. It is, however, not clear how such a geometry would provide the stereoselectivity seen experimentally.

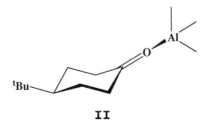

II

We have structurally characterized several organic carbonyl adducts of AlMe(BHT)$_2$. The asymmetric orientation of the aryloxide ligands, due to aluminum-oxygen π-bonding, sterically hinders one face of the coordinated carbonyl (Power, Bott, Atwood & Barron, 1990; Power, Bott, Clark, Atwood & Barron, 1990).

The interaction of AlMe(BHT)$_2$ **(1)** with a series of aldehydes, O=C(H)R, ketones, O=C(R)R', and esters, O=C(OR)R, yields the Lewis acid-base complexes, AlMe(BHT)$_2$[O=C(H)R], AlMe(BHT)$_2$[O=C(R)R'] and AlMe(BHT)$_2$[O=C(OR)R], respectively. The complexes all show a decreased carbonyl stretch, ν (CO), in the IR spectrum and a downfield shift in the ^{13}C-{^{1}H} NMR for the carbonyl α-carbon, when compared to the "free" ligand. These changes are consistent with coordination of the organic-carbonyl to the aluminum center (Power, Bott, Clark, Atwood & Barron, 1990).

The molecular structure of three of these compounds AlMe(BHT)$_2$[O=C(H)^{t}Bu] **(2)**, AlMe(BHT)$_2$(O=CPh$_2$) **(3)** and AlMe(BHT)$_2$[(O=CPh(OMe)] **(4)** are shown in Figures 1, 2 and 3 respectively, along with selected bond lengths and angles. In all three complexes the geometry around aluminum is highly distorted from tetrahedral, with the angles associated with the organo-carbonyl being the most acute. The Al-C bonds are slightly larger than that found for AlMe(BHT)$_2$ [1.927(3)Å] (Shreve, Mulhaupt, Fultz, Calabrese, Robbins & Ittel, 1988). This change is in the direction predicted on the basis of increased p-character in the Al-C bond on changing from planar to pseudo-tetrahedral geometry.

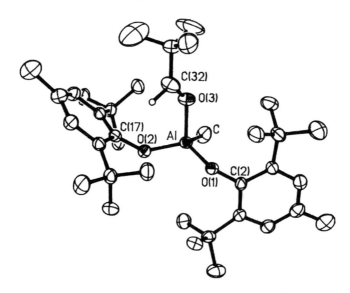

Figure 1. Structure of AlMe(BHT)$_2$[O=C(H)^{t}Bu] (**2**).
Al-O(1) = 1.726(3) Å, Al-O(2) = 1.729(3) Å, Al-O(3)
= 1.920(3) Å, Al-C = 1.955(4) Å, Al-O(1)-C(1) =
131.3(2)°, Al-O(2)-C(11) = 140.5(2)°, Al-O(3)-C(31)
= 136.0(3)°.

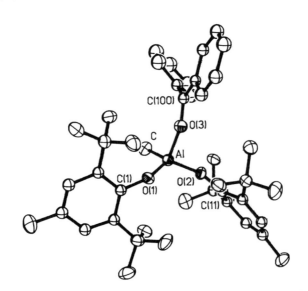

Figure 2. Structure of AlMe(BHT)$_2$(O=CPh$_2$) (**3**). Al-
O(1) = 1.733(5) Å, Al-O(2) = 1.721(6) Å, Al-O(3) =
1.903(6) Å, Al-C = 1.942(8) Å, Al-O(1)-C(1) =
144.0(6)°, Al-O(2)-C(11) = 161.4(5)°, Al-O(3)-C(100)
= 142.6(6)°.

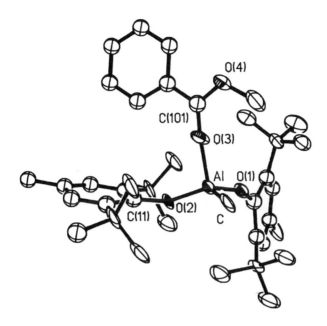

Figure 3. Structure of AlMe(BHT)$_2$[O=CPh(OMe)] (**4**).
Al-O(1) = 1.714(9) Å, Al-O(2) = 1.721(8) Å, Al-O(3)
= 1.851(7) Å, Al-C = 1.96(2) Å, Al-O(1)-C(1) =
153.3(8)°, Al-O(2)-C(11) = 156.3(6)°, Al-O(3)-C(101)
= 174.1(6)°.

The aryloxide Al-O distances [1.714(9) - 1.733(5)
Å] are in the range we have seen before for π-
bonding between oxygen and four-coordinate aluminum
(Healy, Wierda & Barron, 1988; Healy, Ziller &
Barron, 1990). The aryloxide Al-O-C angles in **3** are
different; the larger is comparable to that found
for AlMe$_2$(BHT)(PMe$_3$) (**5**) [164.5 (54)°] (Healy, Wierda
& Barron, 1988) (Figure 4) in which we have shown
significant π-bonding between the oxygen lone pair,
perpendicular to the aryl ring, and the Al-P σ*
orbital. By contrast the second aryloxide has a more
acute Al-O-C angle [142.6(5)°] comparable to that in
AlMe(BHT)$_2$ [140.5(2)°, 146.8(2)°] (Shreve, Mulhaupt,
Fultz, Calabrese, Robbins & Ittel, 1988), where the
lone pairs parallel to the aryl ring are involved in
π-bonding to the vacant aluminum p$_z$ orbital. The Al-
O-C angles in **2** and **4** are similar to that observed
in AlMe$_2$(BHT)py, (**6**) (Figure 5), which has π-bonding
similar to that found in **5** (Healy, Ziller & Barron,
1990).

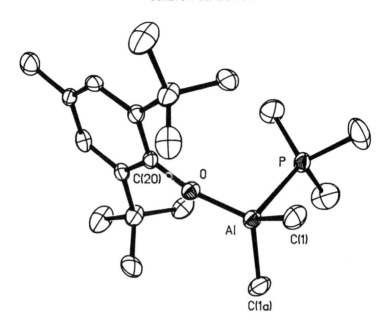

Figure 4. Structure of $AlMe_2(BHT)(PMe_3)$ **5**. Al–O = 1.736(5) Å, Al–P = 2.499(3) Å, Al–C(1) = 1.970(4) Å, Al–O–C(20) = 164.5(4)°.

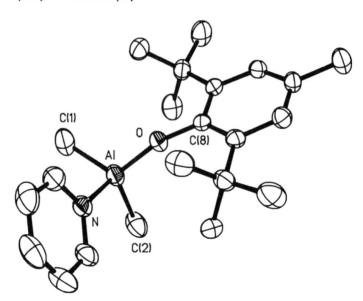

Figure 5. Structure of $AlMe_2(BHT)(py)$ **(6)**. Al–O = 1.740(4) Å, Al–N = 1.993(5) Å, Al–C(1) = 1.964(6) Å, Al–C(2) = 1.956(6) Å, Al–O–C(8) = 156.1(3)°.

On the basis of our structural data we propose that the presence of the π-bonding between the aryloxide lone pairs and the aluminum–carbonyl σ* orbital accounts for the preferred orientation of the two aryloxide ligands.

The presence of the two distinct aryloxide orientations in **2**, **3** and **4** results in the steric protection for one face of the coordinated ligand towards attack from external nucleophilic attack. In the case of tert-butylcyclohexanone (see above) (Maruoka, Itoh, & Yamamoto, 1985; Maruoka, Itoh, Sakuria, & Yamamoto, 1988) the less bulky face of the ketone would be the one hindered by the ortho-tert-butyl groups leaving the sterically more demanding face open for nucleophilic attack (**III**).

III

It is important to note that, although the coordinated aldelyde in **2** and the ketone in **3** are coordinated in an sp^2 manner and the ester in **4** in a linear sp manner, the type of coordination is not a requirement for stereo-selective alkylation. We propose, therefore, that the stereoselectivity seen by Yamamoto and co-workers is controlled by the preferred orientation of the aryloxide ligands due to the aryloxide oxygen–aluminum π-bonding.

Andrew R. Barron

Acknowledgement. The author thanks the International Union of Crystallography for the financial support to attend the Symposium on Organic Crystal Chemistry, Poznan-Rydzyna, Poland.

References

Healy, M.D., Wierda, D.A. and Barron, A.R. (1988) Organometallics **7**, 2543.

Healy, M.D., Ziller, J.W. and A.R. Barron, (1990) J. Am. Chem. Soc. **112**, 2949.

LePage, T.J. & Wiberg, K.B. (1988) J. Am. Chem. Soc. **110**, 6642.

Maruoka, K., Itoh, T. and Yamamoto, H. (1985) J. Am. Chem. Soc. **107**, 4573.

Maruoka, K., Itoh, T., Sakuria, M. and Yamamoto, H. (1988) J. Am. Chem. Soc. **110**, 3588.

Power, M.B., Bott, S.G., Atwood, J.L. and Barron, A.R. (1990) J. Am. Chem. Soc. **112**, 3446.

Power, M.B., Bott, S.G., Clark, D.L., Atwood, J.L. and Barron, A.R. (1990) Organometallics in press.

Shreve, A.P., Mulhaupt, R., Fultz, W., Calabrese, J., Robbins, W. and Ittel, S.D. (1988) Organometallics **7**, 409.

13

Cation–anion interactions in crystalline quininium salts

B. J. Oleksyn, J. Śliwiński, J. Kowalik, and P. Serda

INTRODUCTION

Quinine (Q) is an organic base which contains two nitrogen atoms capable of protonation. One of them, N(1), belongs to the aliphatic quinuclidine moiety which is a tertiary amine, while the other, N(13), is a member of the aromatic quinoline ring. In the environment of polar solvents, protonation of N(1) precedes that of N(13), so that two kinds of quininium cations can be formed:

    i. QH^+ in which only N(1) is protonated,
    ii. QH^{2+} in which both N(1) and N(13) are protonated.

The same is true for three other naturally occurring Cinchona alkaloids which differ from Q in the absolute configuration at C(8) and C(9) and/or in the substituent at C(19), as shown in Fig. 1.

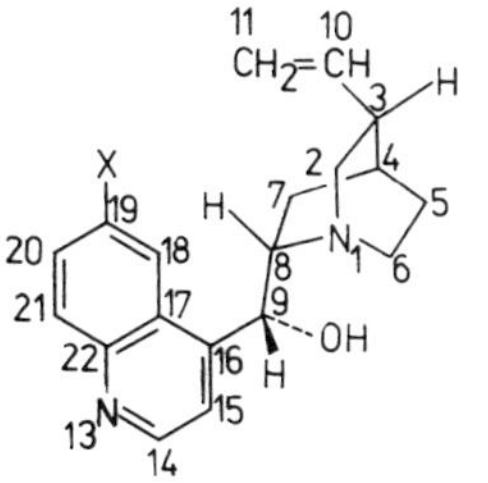

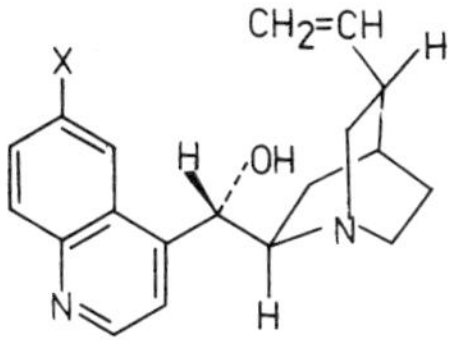

-X	Absolute configuration	Name
-H	C(3)-R,C(4)-S	cinchonidine (I.1)
-OCH₃	C(8)-S,C(9)-R	quinine (II.1)
-H	C(3)-R, C(4)-S	cinchonine (III.1)
-OCH₃	C(8)-R, C(9)-S	quinidine (IV.1)

Fig. 1 Structural formulae and the absolute configuration of Cinchona alkaloids.

The protonated species of Cinchona alkaloids seem to play an important role in the biological activity of these compounds. In particular, the positively charged quinuclidine "head" might be responsible for ionic interactions with active sites of such proteins as muscarinic and α-adrenergic receptors which can be blocked by Cinchona alkaloids. The negatively charged areas of the putative receptor active sites binding QH^+ cations may be modelled by anions which occur in the crystalline structures of the quininium salts. Thus, crystal structure investigation of these salts might provide us with a simplified three-dimensional pattern of a plausible QH^+- receptor interaction.

With this in mind we review here the most interesting aspects of the previously determined crystal structures of the quininium salts and compare them to those of quinine base (Q) and some salts of the Cinchona alkaloids.

Our aim is:
1. to examine changes induced by protonation in the molecular geometry of Q;
2. to discuss cation-anion interactions in the crystalline state, their relationship with the anion type and the number of water molecules in the vicinity of the interacting ions;
3. to consider possible consequences of 1. and 2. for the formation of QH^+-receptor complexes.

The crystal structures considered are listed in Table 1, together with the references to the original structural papers.

MOLECULAR GEOMETRY OF QUININIUM CATION QH^+

Overall shape of the molecules I-IV and of their cations is mainly determined by the torsional angles about two bonds, C(16)-C(9) and C(9)-C(8). The variability of these two angles, τ_1=C(15)-C(16)-C(9)-C(8) and τ_2=O(12)-C(9)-C(8)-C(7), for quininium cations and some other molecules is depicted in Fig. 2, while their values are listed in Table 2.

Table 1. List of the crystal structures considered

SYMBOL	COMPOUND NAME	FORMULA	REFERENCE
I.1	cinchonidine base	$C_{19}H_{22}N_2O$	Oleksyn (1982)
II.1 (Q)	quinine base monohydrate (toluene solvate)	$C_{20}H_{24}N_2O_2 \cdot H_2O \cdot \frac{1}{2} C_7H_8$	Pniewska (1989)
II.1.1	10-hydroxy-10-methyl-10,11-dihydroquinine	$C_{21}H_{28}N_2O_3$	Suszko-Purzycka (1985)
II.1.2	9-acetylquitenine ethyl ester	$C_{29}H_{28}N_2O_5$	Dupont (1985)
II.2	quininium 2,2-dimethyl-cyclopropanecarboxylate monohydrate	$(C_{20}H_{25}N_2O_2)^+ (C_6H_{10}O_2)^- \cdot H_2O$	Graham (1987)
II.3	quininium lactate (orthorhombic)	$(C_{20}H_{25}N_2O_2)^+ (C_3H_5O_3)^-$	Oleksyn (1988)
II.4 A,B	quininium lactate monohydrate (tetragonal)(2 mols in asymm. unit)	$(C_{20}H_{25}N_2O_2)^+ (C_3H_5O_3)^- \cdot H_2O$	Oleksyn (unpub)
II.5	quininium salicylate monohydrate	$(C_{20}H_{25}N_2O_2)^+ (C_7H_5O_3)^- \cdot H_2O$	Oleksyn (1988)
III.1	cinchonine base	$C_{19}H_{22}N_2O$	Oleksyn (1979)

B. J. Oleksyn, J. Śliwiński, J. Kowalik, and P. Serda

Table 1. (continued)

SYMBOL	COMPOUND NAME	FORMULA	REFERENCE
III.2	cinchoninium tetrachlorocadmate dihydrate	$(C_{19}H_{24}N_2O)^{2+}(CdCl_4)^{2-} \cdot 2H_2O$	Oleksyn (1978)
III.3 A,B	bis-[cinchoninium tetrachlorocuprate] trihydrate (2 mols in asymm. unit)	$[(C_{19}H_{24}N_2O)^{2+}(CuCl_4)^{2-}]_2 \cdot 3H_2O$	Dyrek (1987)
III.4	cinchoninium monochloride monohydrate	$(C_{19}H_{23}N_2O)^+ Cl^- \cdot H_2O$	Oleksyn (1987)
IV.1	quinidine base	$C_{20}H_{24}N_2O_2$	Kashino (1983)
IV.1e	quinidine base (ethanol solvate)	$C_{20}H_{24}N_2O_2 \cdot C_2H_6O$	Doherty (1978)
IV.2	quinidinium (−)-1,1'-dimethyl--ferrocene-3-carboxylic acid monohydrate	$(C_{20}H_{25}N_2O_2)^+(C_{13}H_{13}FeO_2)^- \cdot H_2O$	Carter (1967)
IV.3 A, B	quinidinium sulphate dihydrate (2 mols in asymm. unit)	$(C_{20}H_{25}N_2O_2)^+_2 \cdot SO_4^{2-} \cdot H_2O$	Karle (1981)

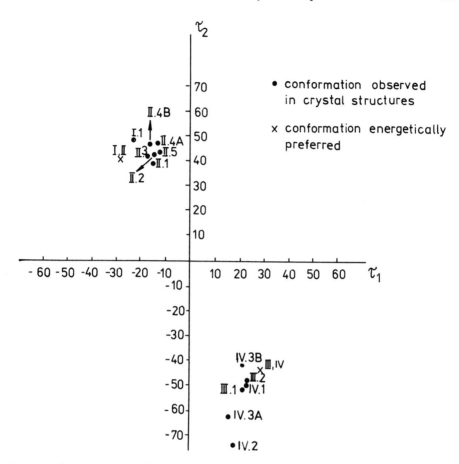

Fig. 2 "Scatterogram" showing the torsion angles
τ_1=C(15)-C(16)-C(9)-O(12) and τ_2=O(12)-C(9)-C(8)-C(7) for some
of the compounds listed in Table 1. Values of τ_1 and τ_2 are
given also in Table 2.

These data reveal the following tendencies in the
conformational behaviour of the respective compounds.

i. The protonation and interactions with anions affect the
conformation of cinchonine (III) and quinidine (IV) to a
greater extent than that of cinchonidine (I) and quinine (II).

ii. Of the two torsion angles τ_2 varies in a wider range
(about 30° for III and IV) than τ_1 (about 15° for III and IV).

Table 2. Quinuclidine geometry and torsion angles about C(9)-C(16) and C(8)-C(9) bonds.

	I.1	II.1	III.1	IV.1	II.2 *)	II.3	II.4 A	II.4 B	II.5	II.2	IV.2	IV.3 *) A	IV.3 *) B
C(2)-N(1)-C(6)	107.3(2)	108.2(3)	107.5(3)	107.5(2)	108.2	108.9(2)	109.9(2)	109.0(5)	109.0(2)	110.3(5)	111(1)	108.1	110.1
C(2)-N(1)-C(8)	107.0(2)	107.8(2)	110.6(3)	111.2(2)	108.5	107.2(2)	108.1(4)	108.6(4)	107.9(2)	112.5(6)	111(1)	113.6	111.4
C(6)-N(1)-C(8)	112.2(2)	110.3(2)	107.1(3)	107.6(2)	113.8	114.5(2)	113.4(4)	112.6(5)	114.2(2)	107.2(5)	104(1)	108.8	108.4
N(1)-C(2)-C(3)	112.3(3)	111.9(3)	112.7(3)	112.4(2)	110.5	110.0(2)	110.6(5)	111.5(5)	109.1(2)	108.3(6)	110(1)	109.7	108.6
N(1)-C(6)-C(5)	112.0(3)	111.7(2)	110.9(3)	110.9(2)	108.8	109.3(3)	107.4(6)	108.4(6)	108.8(2)	106.1(6)	106(2)	108.3	107.4
N(1)-C(8)-C(7)	110.5(2)	111.2(2)	110.1(2)	110.1(2)	108.5	109.3(3)	109.2(5)	108.8(5)	108.0(2)	106.6(5)	108(1)	107.0	107.7
N(1)-C(2)	1.479(4)	1.484(4)	1.474(5)	1.480(4)	1.503	1.506(3)	1.482(8)	1.485(8)	1.497(2)	1.51(1)	1.47(2)	1.511	1.513
N(1)-C(6)	1.477(4)	1.494(3)	1.485(5)	1.474(4)	1.504	1.502(3)	1.506(8)	1.500(8)	1.508(2)	1.50(1)	1.54(3)	1.498	1.509
N(1)-C(8)	1.484(4)	1.471(4)	1.491(4)	1.486(4)	1.516	1.520(3)	1.524(7)	1.510(7)	1.511(2)	1.521(8)	1.55(2)	1.517	1.521
C(15)-C(16)-C(9)-O(12)	-22.6(9)	-14.3(3)	23(1)	23.0(3)	-15.0	-17.36	-14(1)	-16(1)	-11.8(3)	23(1)	18(2)	15.3(7)	20.7(7)
O(12)-C(9)-C(8)-C(7)	48.0(7)	37.9(2)	-51(1)	-50.2(3)	41.7	41.67	47(1)	47(1)	43.7(3)	-48(1)	-74(2)	-62.2(7)	-41.9(7)

*) data on E.S.D. not available

iii. The absolute values of τ_2 rather increase with protonation (except for that of III.2 and IV.3 B) in comparison with the corresponding values for free bases.

The tendency (i) seems to indicate greater flexibility of III and IV in comparison with that of I and II, which might be very interesting in view of differences in biological activity between these two pairs of diastereoisomers. This conclusion needs, however, further evidence, since the compared salts of (II), (III) and (IV) differ in anion types.

The tendencies (ii) and (iii) can be explained by the involvement of the atoms N(1) and O(12) in a hydrogen bonding system with anions, often with the participation of water molecules, which may enforce a more planar arrangement of the atoms N(1), C(8), C(9) and O(12) than in the structures where the alkaloid molecules are hydrogen bonded only to each other.

The geometry of the quinuclidine moiety in the vicinity of N(1) is sensitive to protonation of N(1) which affects both some bond lengths and angles and also conformation, as shown in Table 2 and in Figs. 4a-4f.

In all structures studied, the length of N(1)-C bonds increases on passing from the free base Q(II.1) to the N(1) protonated cations, QH^+ (II.2-II.5). The same observation is valid for cinchoninium and quinidinium salts. A plausible explanation of this lengthening is a modification in the charge distribution which accompanies the involvement of the lone electron pair at N(1) in bonding with proton.

Changes in the bond angles caused by protonation of the quinuclidine fragment are more complex and can be divided into two groups:
(i)   C-N(1)-C angles,
(ii)   N(1)-C-C angles.

(i) In Q(II.1) and in I.1 two of C-N(1)-C angles have values in the range of 107.0(2)-108.2(2)$^\circ$, i.e., are intermediate between the H-N-H angle of ammonia (106.8°) (Dewar 1975) and the tetrahedral angle. The third angle, C(6)-N(1)-C(8), is greater, most probably because of a possible steric interaction between the hydrogen atoms of C(6) and the bulky substituent of C(8). Interestingly, in III.1 and IV.1, where the vinyl group at C(3) and the C(8) substituent are closer to each other than in I.1 and II.1, it is the C(2)-N(1)-C(8) which exceeds the tetrahedral angle, the other two angles being in the range of 107.1(3)-107.6(3)$^\circ$.

The protonation of the free bases cancels the influence of the N(1) lone electron pair on the mutual arrangement of the N(1)- C bonding electrons and leads to an increase in the values of all three C-N(1)-C angles (in 24 out of 27 observations listed in Table 2).

(ii) In Q(II.1) and the other free bases (I.1, III.1 and IV.1) the N(1)-C-C angles are larger than tetrahedral and decrease on protonation; this is consistent with the changes mentioned in (i). A similar behaviour of the quinuclidine moiety on the quaternization of N(1) was noticed previously (Oleksyn 1987).

The conformation of the quinuclidine fragment can be characterized by torsion angles ϕ and ω, shown in Fig. 3. Ermer and Dunitz (1969) derived linear correlations between pairs of these angles in unsubstituted bicyclo[2,2,2]octane which were shown to be approximately valid for the crystalline Cinchona alkaloids, their derivatives and salts (Oleksyn 1987). The points representing Q and QH$^+$ are now added to the diagrams taken from Oleksyn (1987).

The diagrams presented in Figs. 4a-4f reflect an interplay between the steric interactions of substituents at the quinuclidine moiety, in the presence and in the absence of the proton at the N(1) atom, and the influence of crystalline environments.

The points representing pairs of angles, $\phi_1 \omega_2$, $\phi_2 \omega_2'$, and $\phi_3 \omega_2''$ (Figs. 4a-c) observed in the crystalline structures of free bases (I.1, II.1, III.1, IV.1, IV.1e), their derivatives and salts obey, within 3σ values, the Ermer-Dunitz equation:

$$\phi = \frac{3}{5} \omega_2 .$$

An interesting specificity of quinine (II.1), its derivatives (II.1.1, II.1.2) and cation (II.2-II.5) with various anions, is that some of them have negative values of ϕ and ω_2 while only positive angles are observed for cinchonine (III.1), quinidine (IV.1, IV.1e) and their cations (III.2-III.3, IV.2-IV.3). Especially striking is the lack of III and IV structures with angles $\omega_2'' <13°$. The reason for this behaviour of III and IV seems to lie in too short contacts between the substituents at C(8) and C(3) in the case of conformation with negative ϕ and ω_2. From this point of view the angle ω_2'' is particularly important because it involves C(8) with its bulky substituent.

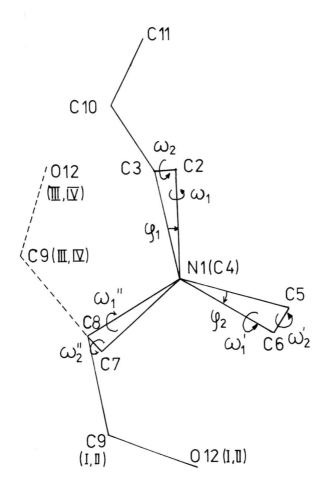

Fig. 3 Quinuclidine moiety projected along N(1)...C(4) line
with the torsion angles:

ϕ_1=C(2)-N(1)...C(4)-C(3) ω_1''=C(7)-C(8)-N(1)-C(6)

ϕ_2=C(6)-N(1)...C(4)-C(5) ω_2=N(1)-C(2)-C(3)-C(4)

ϕ_3=C(8)-N(1)...C(4)-C(7) ω_2'=N(1)-C(6)-C(5)-C(4)

ω_1=C(3)-C(2)-N(1)-C(8) ω_2''=N(1)-C(8)-C(7)-C(4)

ω_1'=C(5)-C(6)-N(1)-C(2)

The orientation of the C(8)-C(9)-O(12) fragment is shown with
solid line for cinchonidine (I) and quinine (II) and with
dashed line for cinchonine (III) and quinidine (IV).

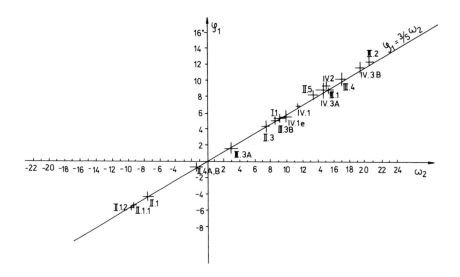

Fig. 4a

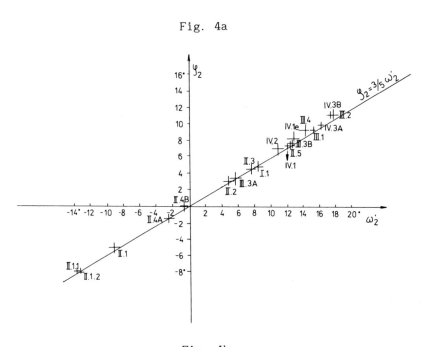

Fig. 4b

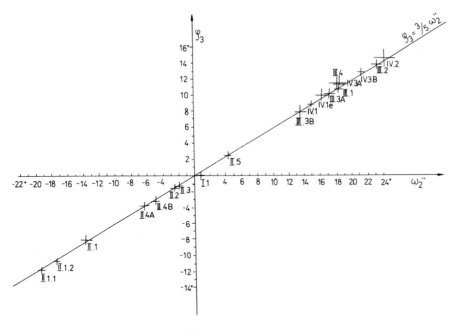

Fig. 4c

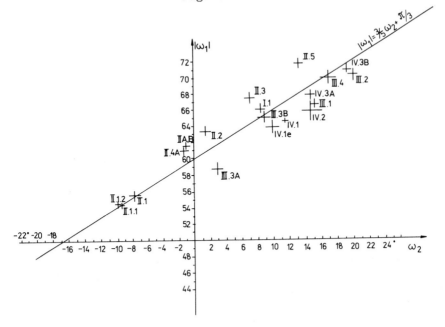

Fig. 4d

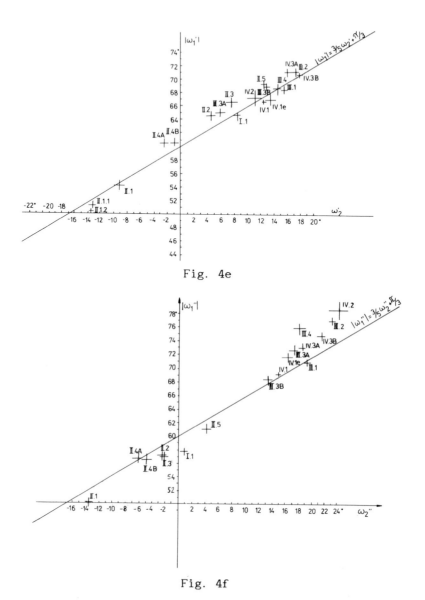

Fig. 4e

Fig. 4f

Fig. 4 Relationships for pairs of quinuclidine torsion angles. The lines represent the theoretical functions postulated by Ermer and Dunitz (1969). Crosses correspond to the observed conformations, cross arms have lengths of 3σ;

a) ϕ_1 vs. ω_2 b) ϕ_2 vs. ω_2' c) ϕ_3 vs. ω_2''
d) $|\omega_1|$ vs. ω_2 e) $|\omega_1'|$ vs. ω_2' f) $|\omega_1''|$ vs. ω_2'' .

This angle in I and II may vary in a wider range than in III and IV, since the S configuration at C(8) excludes such close approach of the two substituents (at C(8) and C(3)) as in the latter two diastereoisomers with R configuration at this atom.

Figs. 4d–4f show that the Ermer–Dunitz equation

$$|\omega_1| = \frac{3}{5}\,\omega_2 + \frac{\pi}{3}$$

does not hold for many of the points, though the tendency of a linear increase of $|\omega_1|$ with ω_2 is preserved. The most significant discrepancy appears in the relationship $|\omega_1''|$ vs. ω_2'' in which the angles are particularly sensitive to the influence of substituents (Fig. 4f).

Protonation of N(1) increases all the angles ϕ and ω leading to a significant twist of the whole quinuclidine fragment. It is also possible that adaptation of the quinuclidine moiety to the intermolecular hydrogen bonding systems, where $-N(1)H^+$ and $-O(12)H$ are proton donors, has some influence on its shape.

CATION–ANION INTERACTIONS

Cation – anion interactions in the salts (II.2 – II.5, III.2, III.3, IV.2, IV.3) of protonated Cinchona alkaloids consist in intermolecular hydrogen bonding in which are involved: protonated N(1) and $-O(12)H$ as donors and anionic oxygen atoms or Cl^- ions as acceptors, often with the participation of water molecules.

In order to understand better the scheme of this bonding, the "mean planes" through the atoms: N(1), H(N1), C(8), C(9), O(12), H(O12), the acceptors of protons and some intermediating atoms, were calculated. The projections on these planes are depicted in Figs. 5a–g, which reveal two kinds of patterns:

(i) closed systems, i.e. 11- or 9-membered rings (abbreviated as HBR), occurring in the structures II.2, II.5, III.2, III.3, IV.2 and IV.3,

(ii) open systems, i.e. infinite chains in II.3, II.4 and IV.3.

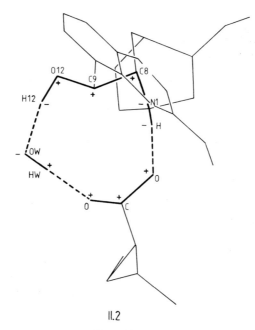

II.2

Fig. 5a

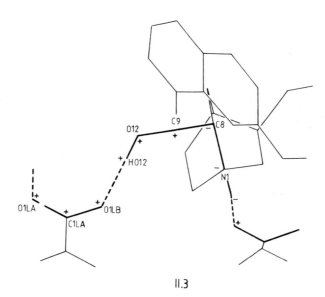

II.3

Fig. 5b

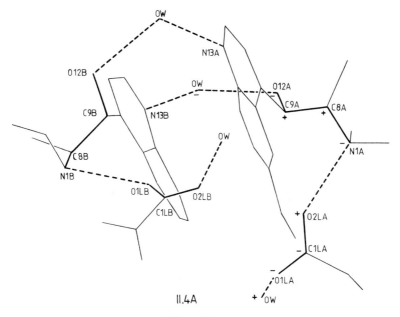

II.4A

Fig. 5c₁

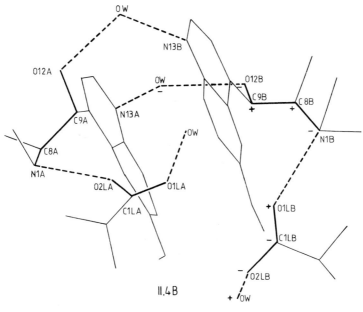

II.4B

Fig. 5c₂

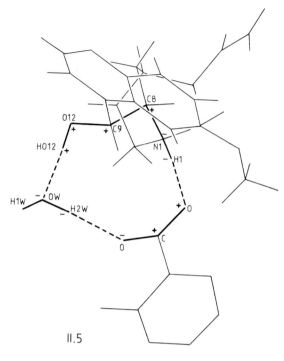

Fig. 5d

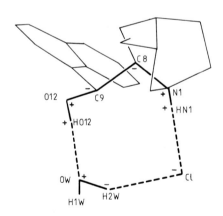

Fig. 5e

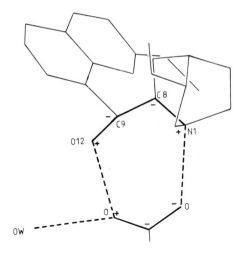

IV.2

Fig. 5f

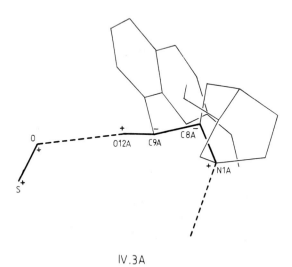

IV.3A

Fig. 5g$_1$

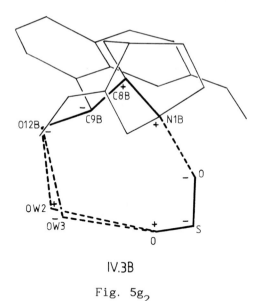

IV.3B

Fig. $5g_2$

Fig. 5 Projections of the studied molecules on the mean planes
through the atoms connected with the thick line. The height of
atoms over and under the planes are marked with plus and
minus, respectively;
a) II.2, b) II.3, c_1) II.4 A, c_2) II.4 B, d) II.5,
e) III.2, f) IV.2, g_1) IV.3 A, g_2) IV.3 B .

In c_1 and c_2 only small fragments of quinuclidine are shown.

(i) In II.2 and II.5 HBR's comprise, besides the H(O12)-O(12)-
C(9)-C(8)-N(1)-H(N1) fragment, also the COO⁻ anion and a water
molecule with one of its protons. The rings have very similar
shapes and dimensions, their orientations with respect to the
quinine molecule being also very comparable.

The same kind of HBR is formed by one of the two quinidine
cations, B, in the asymmetric unit of IV.3 but with the O-S-O
fragment of the sulphate ion in place of the carboxyl. An
interesting modification of this system is an 11-membered ring
in III.3 (not shown here) where two of the four chlorine atoms
bonded to Cu are the acceptors of the protons belonging to the
water molecule linked to O(12) and to N(1).

(ii) Infinite chains of cations connected by anions and/or
water molecules occur only in three of the eight crystalline
salts studied, i.e. in orthorhombic and tetragonal quininium
lactate (II.3, II.4) and in quinidinium sulphate (IV.3). In
II.3 only the carboxylate anions of lactic acid form hydrogen
bonds with QH$^+$ cations of the asymmetric unit in a complex
system not observed in the structures of other salts. In IV.3
the molecules A take part in formation of a chain in which
they alternate with the oxygen atoms of sulphate ions.

The geometry of the atomic groups, ...O...H-O(12)-C(9)-C(8)-
N(1)-H...O..., which are the building blocks of the infinite
chains, is comparable to that of the corresponding fragments
of the closed HBR's. Hence, the reason for the occurrence of
chains seems to lie in the packing conditions of the crystal
structures.

The orientation of the "mean planes" (Figs. 5a–g) with respect
to the quinuclidine and quinoline parts of the alkaloid
cations as well as "folding" of the HBR's or their parts
depend on the absolute configuration of the alkaloids. Only
the distances of the O(12) atom and its proton acceptor from
the mean plane do not change sign with the absolute
configuration, unlike those of C(8), C(9) and N(1).

BIOLOGICAL IMPLICATIONS

The geometrical and bonding features of the Cinchona alkaloid
cations described in this paper should influence their
biological properties. Namely, their conformation, though
rather stable, can adapt to the environmental conditions, the
important torsion angles changing in the range of about 30°.
This flexibility enables the cations to take part in hydrogen
bonding with various anions with or without intermediating
water molecules. The mode of interaction with carboxylic
anions in the crystalline salts may be similar to that
accomplished by Cinchona alkaloids with their protein
receptors, via the side chains aminoacids. In the presence of
water in the vicinity of the receptor site, the water molecule
can assist the aminoacid carboxyl in its interaction with the
alkaloid cation. In more hydrophobic regions the cation can
still interact with the carboxyl but both the cation and the
aminoacid conformations should be modified to achieve the
proper donors–acceptors fitting.

Another important aspect of the anion-cation interaction is
its specificity. As may be noticed in Figs. 5a–g, in quininium
cations(II) the hydrogen-bonded ring (HBR) or its fragment,
together with the quinuclidinium moiety, is always positioned

under the quinoline plane. In contrast, in cinchoninium (III)
and quinidinium (IV) cations, this plane oriented in the same
way (C9-bonded ring on the left side) is placed under the HBR.
This suggests that, if HBR formation is a condition for the
alkaloid cation-receptor complexation, one of the two absolute
configurations, C(9)-R, C(8)-S, or C(9)-S, C(8)-R, should be
preferred to the other. However, in view of the low
specificity of Cinchona alkaloids in their interaction with
some receptors, it is also possible to imagine an alternative
to the above model. Assume that the quinoline moiety is
anchored to a planar hydrophobic region of a putative receptor
site and the HBR is fitted, through conformational
modifications, to form a complex with an aminoacid. Then it is
only the rest of the quinuclidine moiety which will be
differently oriented with respect to the HBR in II as compared
with III and IV cations. This model needs a sufficiently
spacious receptor cavity to host the rather bulky quinuclidine
"head" on either side of the HBR.

More experimental and theoretical investigation is necessary
to verify these tentative considerations.

This work was supported in parts by the World Health
Organization and the Polish Ministry of National Education
Research Project No. RP. II. 13. 2. 13.

REFERENCES

Carter O.L., McPhail A.T., Sim G.A. (1967). *Journal of the
Chemical Society* (A), 365-373.
Ermer O., Dunitz J.D. (1969). *Helvetica Chimica Acta* **52**, 1861-
1886.
Dewar M.J.S., Dougherty R.C. (1975). *"The PMO Theory of
Organic Chemistry"*, New York, Plenum Press.
Doherty R., Benson W.R., Maienthal M., McD. Stewart J.
(1978). *Journal of Pharmaceutical Sciences* **67**, 1698-1701.
Dupont L., Konsur A., Lewiński K., Oleksyn B. (1985). *Acta
Crystallographica* **C41**, 616-619.
Dyrek K., Goslar J., Hodorowicz S.A., Hoffmann S.K.,
Oleksyn B.J., Weselucha-Birczynska A. (1987). *Inorganic
Chemistry* **26**, 1481-1487.
Graham D.W., Ashton W.T., Barash L., Brow J.E., Brown
R.D., Canning L.F., Chen A., Springer J.P., Rogers E.F. (1987).
Journal of Medicinal Chemistry **30**, 1074-1090.
Karle I.L., Karle J. (1981). *Proceedings of the National
Academy of Science USA* **78**, 5938-5941.
Kashino S., Haisa M. (1983). *Acta Crystallographica* **C39**, 310-
312.
Oleksyn B.J. (1982). *Acta Crystallographica* **B38**, 1832-1834.

Oleksyn B. J. (1987). *"Crystal Chemistry of Cinchona Alkaloids and Related Compounds"*, Krakow, Jagiellonian University.

Oleksyn B., Lebioda Ł., Ciechanowicz–Rutkowska M. (1979). *Acta Crystallographica* **B35**, 440–444.

Oleksyn B. J., Pedzińska Z. (1987). *"Molecular Structure of Quinine Derivatives: A Comparative Study"*, (in:) *"Chemistry of Heterocyclic Compounds"*, eds. J. Kovač, P. Zalupsky, Amsterdam, Elsevier, 151–160.

Oleksyn B. J., Serda P. (1988). *Abstracts of International Symposium on Molecular Recognition, Sopron, Hungary*, p. 42.

Oleksyn B. J., Stadnicka K. M., Hodorowicz S. A., (1978). *Acta Crystallographica* **B34**, 811–816.

Oleksyn B. J., Śliwiński J., Kowalik J. (1988). *Abstracts of 2nd International Symposium on Molecular Aspects of Chemotherapy, Gdansk, Poland*, p. 109.

Oleksyn B. J., Śliwiński J., Kowalik J., Dyjak I., Kwoka B., *unpublished data.*

Pniewska B., Suszko–Purzycka A. (1989). *Acta Crystallographica* **C45**, 638–642.

Suszko–Purzycka A., Lipińska T., Piotrowska E., Oleksyn B. J. (1985). *Acta Crystallographica* **C41**, 977–980.

14
Influence of electric field on polymorphism in isotactic polypropylene

D. Paukszta, J. Garbarczyk, and T. Sterzynski

1. INTRODUCTION

As happens with small molecules, polymorphism often also occurs in synthetic polymers. Thus, polyethylene crystallizes in orthorhombic and triclinic forms, polypropylene in monoclinic and orthorhombic forms (Wunderlich 1973; Samuels, Yee 1972; Neapolitano, Pirozzi 1986). These polymers provide examples of this phenomenon and of transformations between the different forms. One among several factors that can lead to the formation of one, rather than another, of the polymorphic modifications is the addition of specific low molecular weight compounds to the polymers.

We are interested in the formation of the hexagonal (β) form of isotactic polypropylene (iPP) (Fig.1) and in its transformation into the more stable monoclinic (α) form (Turner-Jones, Aizlewood, Beckett 1964).

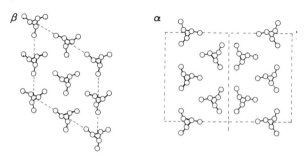

Fig.1 Schematic projections of hexagonal (β) and monoclinic (α) arrangements of iPP helices on planes perpendicular to their axes.

Our earlier investigations and those of other authors have shown that compounds with condensed aromatic rings added to iPP encourage the formation of the β-form (Dragan, Hubney, Muschik 1977; Leugering 1967; Gui-en, Zhi-gun, Jian-min, Zhe-wen 1985; Garbarczyk, Paukszta 1985).

On the grounds of energetic calculations a mechanism of $\beta \rightarrow \alpha$ transition involving several intermediate stages based on rotations and translations of polymer chains was proposed (Garbarczyk 1985) (Fig.2).

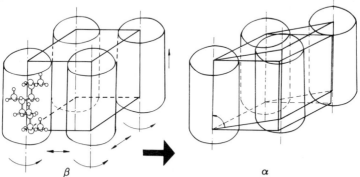

β α

Fig. 2 Schematic illustration of $\beta \rightarrow \alpha$ phase transition, according to the proposed mechanism; the arrows show rotations and translation of helices

According to this hypothesis we considered the role of active additives as retardants of one of the sub-stages of the transformation. The suggested mechanism was confirmed partially by using wide angle X-ray scattering (WAXS) as well as synchrotron radiation methods (Garbarczyk, Paukszta 1985; Garbarczyk, Paukszta, Sterzynski 1989). However, the problem of direct experimental confirmation of the hypothesis on the mechanism of interaction between additive and polymer chains still remains unsolved. There are essentially two reasons which make it difficult to prove the exact mechanism : the first is the coexistence of amorphous and crystalline phases in the solid polymer and the second is the small concentration of additives in analysed materials. Therefore, it is necessary to apply several methods which can provide only an approach to the solving of the problem.

Our previous observations have suggested that the action of the compounds in question consists in the formation of molecular complexes of an additive with the polymer chain or in interactions between the surface of the crystal of additive and the polymer. It is well known that such weak interactions are sensitive to electric field (EF); therefore we suspected, that as a result of EF action changes in the amount of β phase as well as in the kinetics of the $\beta \rightarrow \alpha$ phase transition should be observed.

We present below some results of a study undertaken to confirm the above hypothesis.

The samples for the investigations were prepared by non-isothermal crystallization of melted iPP as previously described (Garbarczyk, Paukszta 1981). We examined monoclinic isotactic polypropylene (without additives) as well as samples of iPP containing both α and β forms. The hexagonal polymorphic modification was obtained in the presence of two β nucleators : Permanent Red E3B (Hoechst, FRG) or triphenodithiazine (TPDT) (Garbarczyk, Zuk 1979). Two kinds of experiments have been performed :

Test I. Crystallization in an electric field

The samples of iPP containing 0.5% of E3B were heated up to 488K (40K above melting temperature), kept at this temperature for 15 minutes and subsequently crystallized under non-isothermal conditions. The electric field was applied before melting of the polymer (at 423K) and switched off after crystallization (at 333K). A diagram of our apparatus is shown in Fig. 3.

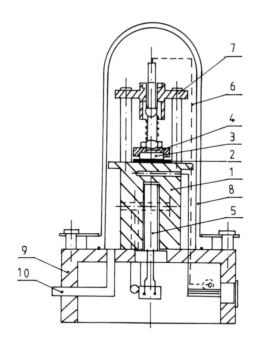

Fig. 3. Apparatus for applying the EF : 1 - electrode with heater, 2 - sample, 3 - upper electrode, 4 - electrode handle, 5 -heater, 6 -high current supply, 7 - handle and isolation, 8 -glass cover, 9 -base plate, 10 -dried nitrogen.

Test II. Phase transition in electric field.

The samples for this test were prepared by crystallization of iPP with E3B without EF. The phase transition was performed by heating up to 430K and EF was supplied at 423K. All the experiments of tests I and II were carried out according to the temperature and voltage program shown in Figs. 4a and 4b, respectively.

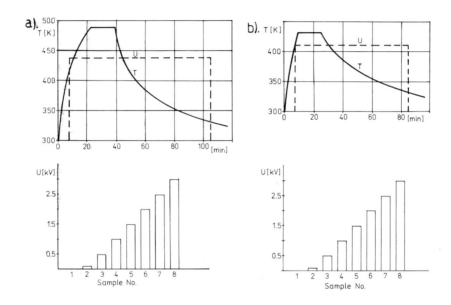

Fig.4 Temperature and voltage during test I a). and test II b).; the lower figures show the applied voltage during the study.

Structural investigations were carried out at room temperature by means of X-ray diffraction on a semi-automatic horizontal diffractometer, using CuK_α radiation. The amount of β form in the iPP samples (k) was determined on the basis of the intensities of the $(110)_\alpha$, $(040)_\alpha$, $(111)_\alpha$ and $(300)_\beta$ reflections by use of the Turner-Jones (1964) formula. We analyzed the relative changes of k values: $k_v/k_{ov} \, 10^2$, where k_v is the amount of β form in samples at given voltage and k_{ov} determines the amount of β phase in samples crystallized under the same thermal conditions but without EF.

2. RESULTS AND DISCUSSION

The results from test I (Fig.5) showed a very high influence of the EF on the course of crystallization of polymer containing active additive. Under these conditions we observed that the formation of the monoclinic form is strongly preferred.

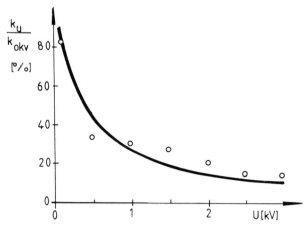

Fig. 5. Relative changes in the amount of β phase for sample iPP with E3B as a function of voltage applied during crystallization.

Up to 0.1 kV a small decrease of the amount of the β form was observed (k_v/k_{ov} = 18%), but above 0.5 kV the decay of this form increased considerably. For 2.5 kV the relative amount of the β phase drops below 80%. With further increase of the voltage, traces of β phase were still observed on the X-ray diffractograms. These were detected by small but significant maxima on the diffraction patterns at angles $\theta = 8.06°$ and $\theta = 10.57°$.

According to the mechanism mentioned above, the formation of the β form is connected with the decrease in the mobility of macromolecular chains during the cooling process of molten iPP, which depends on the strength of the interaction between additive and polymer. We assume that the electric field causes disturbances in these interactions, which are weaker and, therefore, the spatial arrangement of iPP may be easily changed. Disturbances may occur in the electron level in the additive as well as in the polymer and thus the EF makes the London interactions much weaker. The second possible explanation of the observed phenomenon involves a tendency for an orientation of polymer chains by the EF which then makes it easier for the macromolecules to translate parallel to the axes of the helices. According to the mechanism described previously (Garbarczyk 1985) the creation of the a modification requires such translations. But analysis of the X-ray diffraction pattern showed that after electrocrystallization the orientation of iPP crystallites remained unchanged. Consequently, the above hypothesis must be rejected.

We must also consider an orientation of the small molecules caused by EF. Although E3B has oxygen and nitrogen atoms, which may cause stable displacement of electric charge, the symmetrical arrangement of these atoms in the molecule (Geissler 1977) excludes this effect.

In our previous study we found that the disappearance of the hexagonal modification during heating of β-iPP is equivalent to transformation of this modification into the α phase (Garbarczyk, Paukszta 1985, 1989). Therefore changes in the amount of the β phase determine the kinetics of the $\beta \rightarrow \alpha$ transition. The β phase does not disappear at a strictly determined temperature but it happens within a temperature range which depends on the kind of additive added. The temperature chosen for the test II was below the temperature of the full phase transition. The results of test II indicated that the course of phase transition in question is also sensitive to the EF (Fig. 6), but less so than the crystallization process.

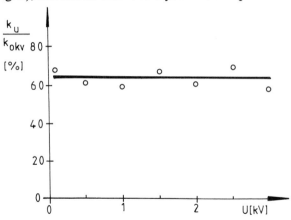

Fig. 6. The course of the $\beta \rightarrow \alpha$ phase transition at 430K as a function of voltage; shows the relative value of k after heating without EF.

In the case of heating without EF the decrease in the amount of β-iPP was less than 20%. However, a weak EF (100 V) caused decay of this phase up to 67%. But it must be emphasized that during further increases of voltage the amount of β phase remains unchanged. Therefore the present experiments unequivocally show that below the temperature of the full transition the EF influences this process only to a limited extent. It accords with our earlier work, where we found that the essential factor causing the disappearance of the β modification is the temperature and not the time of the thermal treatment (Garbarczyk, Paukszta 1985).

In order to test the hypothesis that a strong EF diversified the samples with regard to ability to polarization, the investigations of electric permittivity (ε) were carried out. Samples of iPP crystallized in the EF (at 2.5 kV) and without EF were tested by means of a capacitance bridge (Hewlett Packard HP 4270A) at frequency 10 kHz and at series of temperatures from 293 up to 453K.

The results of this study formulated as $\Delta\varepsilon/\varepsilon_{293}$ as a function of temperature are shown in Fig.7, where $\Delta\varepsilon$ is the difference between ε at a given temperature and at 293K.

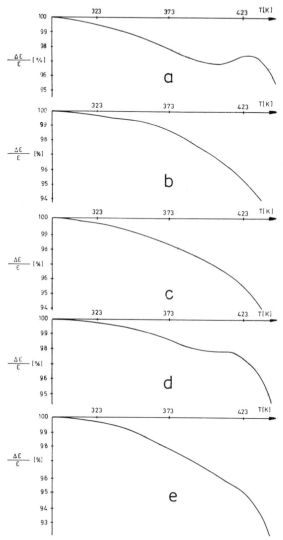

Fig. 7. Relative changes of electric permittivity ($\Delta\varepsilon/\varepsilon_{293}$) vs. temperature for the following samples : iPP with E3B crystallized without EF a). and in EF b).; pure iPP crystallized in EF as well as without EF (the same curve) c)., iPP without TPDT crystallized without EF d). and in EF e).

As had been expected, due to thermal expansion the relative values of electric permittivity ($\Delta\varepsilon/\varepsilon$) for all samples decrease with the temperature. However, the character of the curves for iPP (with E3B) crystallized without EF differs appreciably from those for samples crystallized in EF. It should be mentioned that transformation from the non-centrosymmetric polar β form (space group $P3_1$ or $P3_2$) (Turner-Jones, Aizlewood, Beckett 1964; Geil 1963) to the centrosymmetric nonpolar α form (space group $P2_1/c$) (Geil 1963, Corradini, Petracone, Pirozzi 1983) strengthened the tendency of ε to decrease with decrease of temperature.

The anomaly (423K) (Fig.7a) for samples containing 62% of β phase is connected with the $\beta \rightarrow \alpha$ transformation. The last stage of transition by the proposed mechanism consists in a translation of macromolecules perpendicular to their axes. As a consequence a decrease in lattice volume should take place which causes an increase of ε.

For verification of this suggestion we tested changes of ε in samples containing another β-nucleator (TPDT) with a very similar chemical structure to the E3B (Garbarczyk 1985, Acta Cryst.). The observed plateau (Fig. 7d) for this sample crystallized without EF (k=48%) confirmed the above observation and thus showed the possibility of monitoring the $\beta \rightarrow \alpha$ transition by means of ε measurements.

It is interesting that, in the case of samples containing mainly monoclinic iPP, the function $\Delta\varepsilon/\varepsilon$ vs. temperature is almost linear (Fig. 7b, 7c, 7e).

3. CONCLUSIONS

I. A strong electric field significantly influences crystallization of isotactic polypropylene containing structurally active additives. Under such conditions much less hexagonal form arises, which indicates a weakening of action of β-nucleators

II. The electric field is a factor accelerating the $\beta \rightarrow \alpha$ phase transition but the main role in the transformation is played by the temperature.

III. Changes in electric permittivity during annealing also confirm the theoretically expected mechanism of polymorphic transformation.

4. ACKNOWLEDGEMENTS

The authors gratefully acknowledge Mr.J.Wolak (Institute of Molecular Physics, Polish Academy of Sciences, Poznań) for help during the measurements of electric permittivity. This work was supported by the Polish Academy of Sciences, Project No. C.P.B.P.01.12.

5. REFERENCES

Corradini P., Petracone V., Pirozzi B., (1983), Eur. Polym.J., 19, 299.

Dragan H., Hubney H., Muschik M., (1977), J.Pol.Sci.Phys. Ed.15, 1, 1779.

Garbarczyk J., Zuk A., (1979), Phosphur and Sulfur, 6, 351.

Garbarczyk J., Paukszta D., (1981), Polymer 22, 562.

Garbarczyk J., (1985), J.Makromol.Sci., 186, 145.

Garbarczyk J., Paukszta D., (1985), Coll.Polym.Sci., 263, 985.

Garbarczyk J., (1985), Acta Cryst., C41, 1062.

Garbarczyk J., Sterzynski T., Paukszta D., (1989), Polym.Commun., 30, 153.

Geil P.H., (1963), Polymer Single Crystals, J.Wiley & Sons, New York.

Geissler G., (1977), Dtsch. Farben-Z., 31, 190.

Gui-en Z., Zhi-gun H., Jian-min Y., Zhe-wen H., Guan-g S., (1985), Makromol. Sci., 263, 985.

Leugering H.J., (1967), Makromol.Chem., 109, 204.

Neapolitano R., Pirozzi B., (1986), Makromol.Chem. 187, 1993.

Samuels R.J., Yee R.Y., (1972), J.Polym.Sci., A2, 10, 385.

Turner-Jones A., Aizlewood J.M., Beckett D.R., (1964), Makromol.Chem., 75, 134.

Wunderlich E., (1973), Macromolecular Physics, Acad.Press, New York, Vol.1.

15

Organic crystal chemistry in the 1990's

D. W. Jones

In the final session of the Symposium, A. Kálmán (Hungary) chaired a round-table session in which J. Bernstein (Israel), R. Boese (F.R. Germany), L. Leiserowitz (Israel) and J. Lipkowski (Poland) voiced personal thoughts and predictions about the shape of organic crystal chemistry in the next few years. From these individual contributions, one could discern four main themes, each raised or inferred by more than one speaker: (1) collaborative approaches in structural science; (2) use of accumulated crystallographic data; (3) emerging topics in crystal chemistry; and (4) exploitation of newer techniques.

(1) Collaborative approaches to structural science

In many countries over the past two decades, crystallographers have (except in a few essentially crystallographic laboratories) often become dispersed into very small entities within a department concerned with some branches of main-stream chemistry. Crystal structure determination is often a service activity or may be performed (no doubt usefully but, from a crystallographic viewpoint, imperfectly) by a non-crystallographer, whose main expertise might be in organic synthesis; significant discrepancies may be overlooked by the "amateur crystallographer". Leiserowitz recalled Prelog's comparison of chemists as inventors, whose first aim is to make compounds, with biologists who may be regarded more as historians or discoverers, whose aim is understanding. Organic crystal chemistry can involve the collaboration of organic, physical and theoretical chemists and biologists, as well as X-ray and neutron crystallographers. Kálmán felt that genuine partnerships between organic chemists and crystallographers were desirable in furthering structural science. This could only be achieved by the better articulation of structural views in joint programmes which, in appropriate cases, would also incorporate molecular spectroscopy and quantum chemistry calculations.

(2) Use of accumulated crystallographic data

Through the pioneering Cambridge Crystallographic Data Base, the near explosive growth in reliable crystallographic data, made possible by direct methods and computing power in structure solution and refinement, is now widely accessible. Both Bernstein and Kálmán were anxious that this rich source of structural information should be fully utilized in the study of the systematics of molecular packing, conformation, biological action and structure-activity relationships. Chemists should be involved in the statistical analysis of molecular and crystal structures. There is also the possibility that the essential crystallographic data from many supposedly routine analyses may go unrecorded with merely a structural result appearing in an essentially organic chemical paper.

(3) Emerging topics in crystal chemistry
As Bernstein has emphasised, polymorphism seems to be so widespread and significant (even if this is not fully realized) that the interplay between crystal forces and molecular conformation is a fertile area for study. Related to polymorphism are problems raised by recent structural data on organic solvates, intercalates, inclusion compounds, organic clays, zeolites and clathrates-structures which cannot be interpreted in terms of the close packing of the host component. Considering that the Kitaigorodsky model is appropriate only to close packed, or rather stable, molecular structures, Lipkowski stresses the need for a new approach to crystal packing in such multi-component molecular crystals. More generally, structure-property relationships in these systems present a formidable challenge for joint attack by crystallographers, chemists and biologists. The structure and chemistry of layers and surfaces will be more open to study by X-ray crystallographers through the application of synchrotron radiation. The biological action of organic molecules and their interaction with proteins will continue to be a major concern of crystallographers collaborating with biologists, biochemists, etc. Other emerging areas highlighted by Bernstein for crystallographic study in the near future included new materials such as those involved in magnetics, molecular electronics, inclusion compounds and extended structures.

(4) Exploitation of newer techniques
Although the overall merits of determining crystal structures at low temperature for coordinate accuracy (or time saving) have long been clear, a remarkably small proportion of analyses is actually performed at other than room temperature. As an advocate of determining any crystal structure at low temperature, Boese pointed out that, aside from detection of solid-state phase transitions or disorder (and possibly avoiding the latter), low-temperature single-crystal X-ray crystallography helps in any or all of direct-methods structure solution, hydrogen location, resolution of rigid/non-rigid body motions, analysis of dynamical effects and study of electron density distribution. Obviously heat-sensitive low-melting compounds must have low temperature study. For the design and construction commercially of crystal-friendly (no ice), budget-friendly ($1\ 1\ N_2\ h^{-1}$ gas consumption in aerodynamical stream), computer-controlled, well-insulated cryostats operating at a stable 100 K or less at the crystal, Boese appealed to manufacturers to make full use of customer experience.

Similarly, the deformation density approach, though by no means new, was one that, in Leiserowitz's opinion, should be more frequently employed.

Neither synchrotron X-ray nor high-flux continuous or pulsed (e.g. SNS) neutron sources are widely available in Eastern Europe but both are very relevant to organic crystal chemistry in the 1990's and both kinds of source are accessible to collaborative or visitor usage.

Finally, in the light of remarkable political changes begun in late 1989, it is appropriate to mention Kálmán's concern that help from more developed laboratories to Eastern European colleagues in the application of molecular graphics should not be inhibited by political limitations (the COCOM list) on computing facilities.